TOPOLOGY OF SURFACES

By ANDRÉ GRAMAIN

Originally published as TOPOLOGIE DES SURFACES
by André Gramain

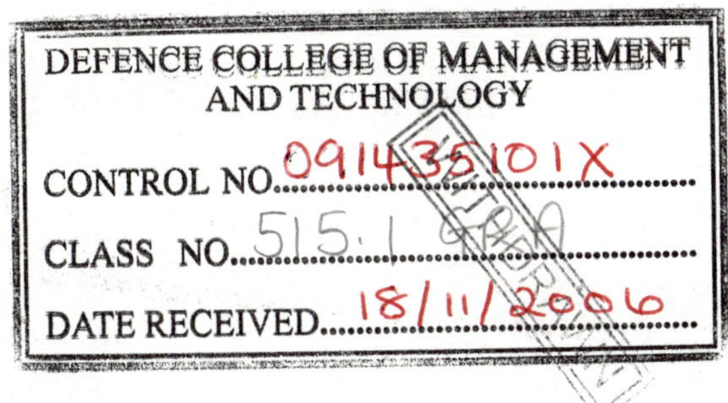

Translated from the French and edited by
LEO F. BORON, CHARLES O. CHRISTENSON and
BRYAN A. SMITH, all of the University of
Idaho, Moscow, Idaho, USA.

BCS ASSOCIATES
Moscow, Idaho, U.S.A.
1984

Manufactured in the United States of America

ISBN: 0-914351-01-X

Library of Congress Catalog Card Number: 84-70461

CONTENTS

CHAPTER III

DIFFERENTIABLE FUNCTIONS AND MANIFOLDS

CHAPTER IV

MORSE FUNCTIONS ON SURFACES

CHAPTER V

THE CLASSIFICATION OF SURFACES

CHAPTER VI

KNOTS

CHAPTER VII

SURFACES IN EUCLIDEAN SPACE R^3

FOREWORD TO THE FRENCH EDITION

This book is the edited version of lecture notes used for a course in the Faculté des Sciences d'Orsay during the academic year 1969-1970. The course had the twofold goal of showing students several methods of algebraic and differential topology and of proving some elementary results of topology. To the first objective there corresponds the study of the notions of the fundamental group of a space and of the Morse function on a manifold; to the second objective, the proof of the theorem on the classification of compact surfaces, using the preceding notions.

The mathematical background needed to understand this book is modest. An undergraduate course in topology will suffice. Thus, we have not discussed topological spaces but only metric spaces, and (except in Chapter VII) we have replaced the concept of connectedness by that of arcwise connectedness, and manifolds are always submanifolds of a Euclidean space. True, this deliberate choice of this level has some shortcomings. We had to assume the existence of a Morse function on a manifold and the classification theorem, up to diffeomorphism, of compact differentiable surfaces is not proved, nor is the theorem on the classification of "topological" surfaces. We content ourselves with the classification up to homeomorphism of differentiable surfaces to avoid problems concerning the gluing together of diffeomorphisms.

Chapters I and II introduce the notion of the fundamental group and give examples of the calculation of it using the Van Kampen theorem, notably for compact surfaces.

Chapters III and IV are devoted to the definition and study of manifolds and Morse functions on manifolds. The existence of Morse functions is the only "global" theorem that we had to assume.

The classification of surfaces is treated in Chapter V. This is an example of the use of Morse functions. The technique of gluing together homeomorphisms is a little tedious, but it is more susceptible to generalization than the classical combinatorial methods.

The last chapter, Chapter VII, contains an "elementary" proof of the theorem on the decomposition of Euclidean space into two components by a compact connected surface. This proof does not make use of homology theory. Alexander's duality was avoided by judicious use of the fundamental group of a surface and the group of a knot. That is why Chapter VI furnishes some elements of knot theory.

The reader who desires to go more deeply into the topics treated in this book and still remain at a rather elementary level can read:

M. GREENBERG, Lectures on algebraic topology, Benjamin, Inc., (1967) for algebraic topology;

J. MILNOR, Topology from the differentiable viewpoint, The University Press of Virginia (1965) for differential topology.

ANDRÉ GRAMAIN

Paris, France

1970

FOREWORD TO THE ENGLISH EDITION

The present English edition is a faithful translation of the original edition of 1971 with one exception. Section 1 of Chapter VI of the French edition contained an error. In this edition, I have pointed out this error and of course I have redone this part.

As possible further reading for the young student who would like to become further acquainted with topology, still at an elementary level, I add the classical book:

SEIFERT AND THRELFALL, A Textbook of Topology, Academic Press (1980).

ANDRÉ GRAMAIN

Tours (France)

1984

CHAPTER I

THE FUNDAMENTAL GROUP

We will associate with each arcwise connected metric space X a group $\pi(X)$ and with each continuous mapping f: $X \to Y$ a homomorphism $\pi(f):\pi(X) \to \pi(Y)$. Two homeomorphic[1] spaces X and X' will have isomorphic fundamental groups $\pi(X)$ and $\pi(X')$, respectively. Hence, the group $\pi(X)$ reflects some of the topological properties of the space X. We shall work toward obtaining the groups $\pi(X)$ of some well-known spaces X.

§1. ARCWISE CONNECTEDNESS

Let X be a metric space. A *path* in X is a continuous mapping γ: $[a,b] \to X$ where a, $b \in R$, $a < b$. In general, we may assume that $a = 0$ and $b = 1$ since this case can always be obtained by using the unique linear affine mapping of $[a,b]$ onto $[0,1] = I$. We call the point $\gamma(0)$ the *initial point* and the point $\gamma(1)$ the *terminal point* of the path γ. The space X is called *arcwise connected* if any two points x, $y \in X$ can be joined by a path having initial point x and terminal point y.

Example 1. Every convex subset of R^n (in particular, every ball) is arcwise connected.

---

[1]We say that X and X' are *homeomorphic* if there exists a bijective (i.e., one to one and onto) mapping f: $X \to X'$ which is continuous, as is its inverse. In this context f is called a *homeomorphism*.

PROPOSITION 1. *The union of a family* $\{A_\beta\}_{\beta \in B}$ *of arcwise connected subsets of* X, *having a point* p *in common, is arcwise connected.*

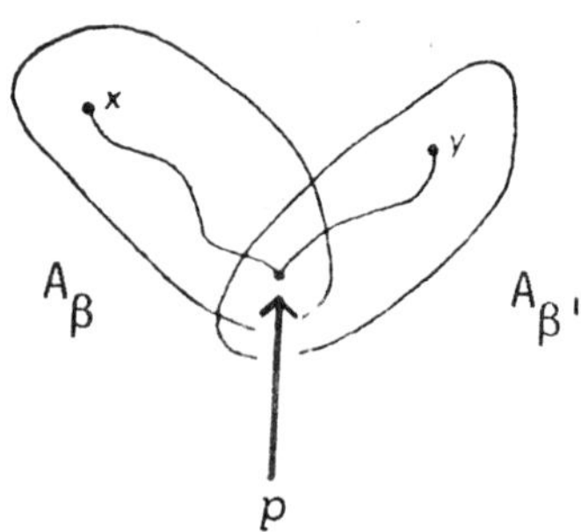

If $x \in A_\beta$ and $y \in A_{\beta'}$, we obtain a path from x to y by joining a path from x to p with a path from p to y.

PROPOSITION 2. *The relation* $R(x,y)$ = "there exists a path that joins x to y" *is an equivalence relation. The equivalence class* (i.e., the arcwise connected component) *of a point* x *is the largest arcwise connected subset containing* x.

The proof is easy.

PROPOSITION 3. *The continuous image of an arcwise connected space is arcwise connected.*

In fact, if γ is a path that joins x to y in X, then the path $f \circ \gamma$ joins $f(x)$ to $f(y)$ in $f(X)$.

We say that a space X is *locally arcwise connected* if every point x of X has a neighborhood basis consisting of arcwise connected sets. For example, an open subset of R^n is locally arcwise connected, and more generally, so is a topological manifold.[1]

---

[1]A metric space X is called an *n-dimensional topological manifold* if every point of X has a neighborhood that is homeomorphic to R^n.

PROPOSITION 4. *In a locally arcwise connected space, the arcwise connected components are open and closed.*

It follows from Propositions 1 and 2 that an arcwise connected component is open. An arcwise connected component is closed since its complement, being the union of the remaining components, is open.

Remark 1 (connected spaces). A connected space is a space which does not have any subset other than the empty set and itself which is simultaneously open and closed. It follows from the preceding proposition that a connected, locally arcwise connected space is also arcwise connected, inasmuch as there is only one component. We note that every arcwise connected space is connected. This results from the fact that the segment [0,1] and hence the image of every path is connected.

§2. PATHS: EQUIVALENCE, PRODUCTS

Equivalence relation. We say that two paths γ_0 and γ_1 in X are *equivalent* (or *homotopic* with fixed endpoints), and we denote this by $\gamma_0 \sim \gamma_1$, if there exists a continuous mapping $\delta: I \times I \to X$ such that:

$$\delta(t,0) = \gamma_0(t),\ \delta(t,1) = \gamma_1(t), \quad \text{for } 0 \leqq t \leqq 1;$$
$$\delta(0,s) = \gamma_0(0) = \gamma_1(0),$$
$$\delta(1,s) = \gamma_0(1) = \gamma_1(1), \qquad \text{for } 0 \leqq s \leqq 1.$$

Two equivalent paths have the same initial point x_0 and the same terminal point x_1. Intuitively, if we set $\gamma_s(t) = \delta(t,s)$, then the family of paths γ_s, each having initial point x_0 and terminal point x_1, is a *continuous deformation* of γ_0 into γ_1 as s varies from 0 to 1. To be precise, the mapping $s \mapsto \gamma_s$ is a continuous path in the space of paths (with the topology of uniform convergence) having initial point x_0 and terminal point x_1.[1]

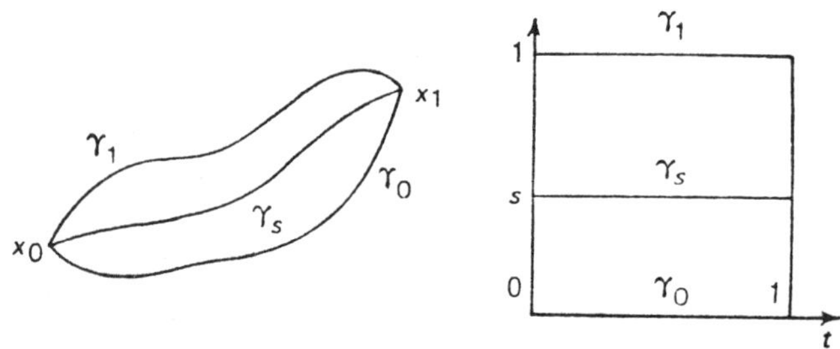

Product of paths. Intuitively, we are dealing here with the operation of joining two paths (already used in §1). Let γ' and γ'' be two paths in X such that the initial point $\gamma''(0)$ coincides with the terminal point $\gamma'(1)$. The product $\gamma = \gamma' \cdot \gamma''$ is defined by the following formula:

$$\gamma(t) = \begin{cases} \gamma'(2t) & \text{if } 0 \leqq t \leqq \frac{1}{2}, \\ \gamma''(2t - 1) & \text{if } \frac{1}{2} \leqq t \leqq 1. \end{cases}$$

PROPOSITION 5. *The equivalence class of the product path depends only on the classes of the factor paths.*

Let γ_0' and γ_0'' be paths such that $\gamma_0'(1) = \gamma_0''(0)$. Let δ' be a homotopy between γ_0' and γ_1' and let δ'' be a homotopy between γ_0'' and γ_1''. Then γ_1' and γ_1'' may be multiplied and we have

$$\gamma_0' \cdot \gamma_0'' \sim \gamma_1' \cdot \gamma_1''.$$

---

[1]We shall show that, for each $\varepsilon > 0$, there is an α such that, if $|s - s_0| \leqq \alpha$, then $\sup_{t \in I} d(\gamma_s(t), \gamma_{s_0}(t)) \leqq \varepsilon$. Suppose not; then there exists an ε and for each $n \geqq 0$ a point $(t_n, s_n) \in I \times I$ such that $|s_n - s_0| \leqq \frac{1}{n}$ and $d(\gamma_{s_n}(t_n), \gamma_{s_0}(t_n)) > \varepsilon$. This sequence $\{(t_n, s_n)\}$ has an accumulation point (t_0, s_0). The continuity of δ at (t_0, s_0) yields the contradiction.

In fact, we set

$$\delta(t,s) = \begin{cases} \delta'(2t,s) & \text{if } 0 \leqq t \leqq \frac{1}{2}, \\ \delta''(2t-1,s) & \text{if } \frac{1}{2} \leqq t \leqq 1. \end{cases}$$

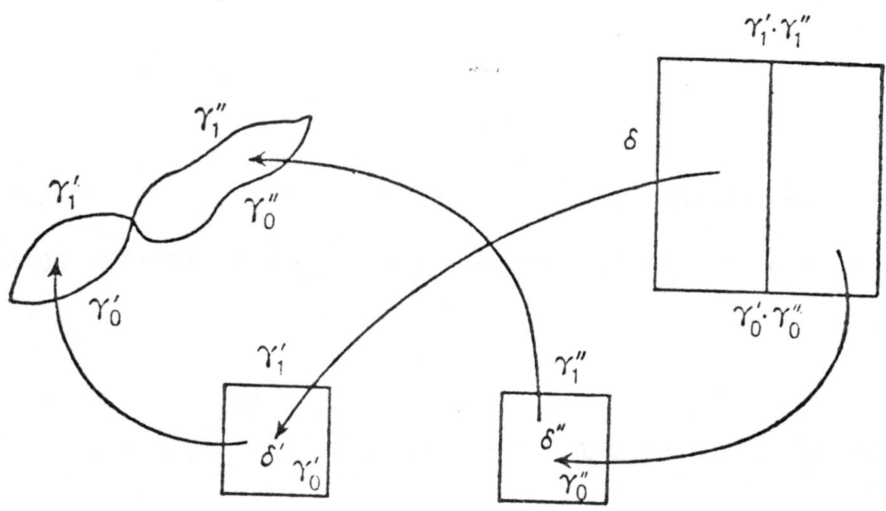

The mapping δ is a homotopy between $\gamma_0' \cdot \gamma_0''$ and $\gamma_1 \cdot \gamma_1''$. To verify that δ is continuous (as well as that $\gamma' \cdot \gamma''$ defined above is continuous) we use the following result.

PROPOSITION 6. *Let f: X → Y be a well-defined mapping and let* A *and* B *be two closed subsets of* X *such that* X = A ∪ B. *If the restrictions $f|$A and $f|$B are continuous, then f is continuous.*

Associativity. The product is associative up to equivalence class; that is,

$$\gamma\cdot(\gamma'\cdot\gamma'') \sim (\gamma\cdot\gamma')\cdot\gamma''$$

when the products are defined. An explicit formula for this homotopy can be obtained by considering the following figure.

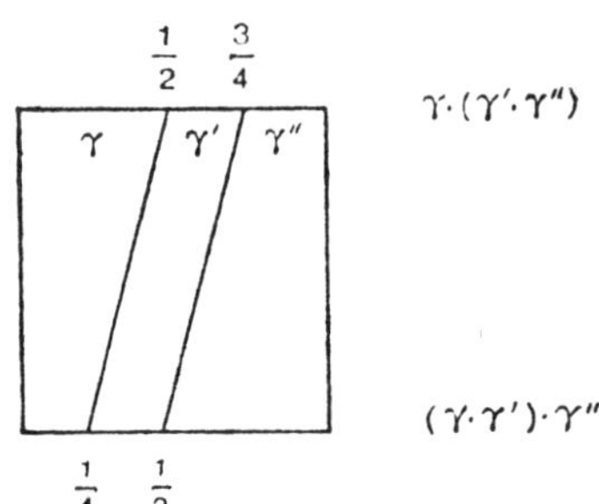

Neutral and inverse elements. We denote by c_x the *constant* path whose image is $x \in X$. For every path γ having initial point x and terminal point y we have:

$$c_x \cdot \gamma \sim \gamma \quad \text{and} \quad \gamma \cdot c_y \sim \gamma.$$

We indicate this with the aid of the following drawings:

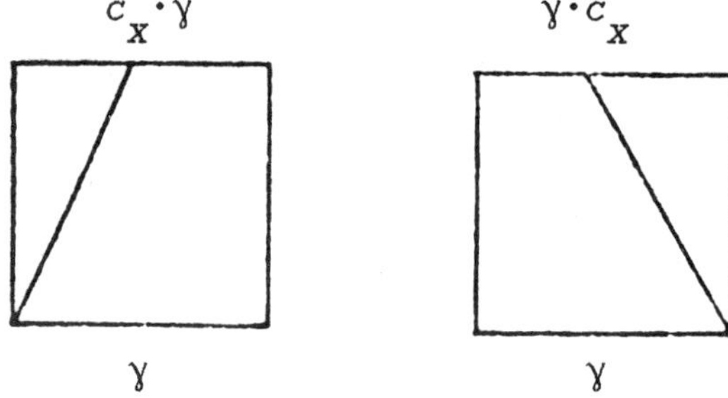

Let γ^{-1} denote the path defined by $\gamma^{-1}(t) = \gamma(1 - t)$. Then we have

$$\gamma \cdot \gamma^{-1} \sim c_x \quad \text{and} \quad \gamma^{-1} \cdot \gamma \sim c_y.$$

We say that γ^{-1} is the *inverse* of γ.

§3. LOOPS, THE FUNDAMENTAL GROUP

A path γ in the space X is called a *loop* at $x \in X$ (or, a *loop with base point* x) if its initial point $\gamma(0)$ and its terminal point $\gamma(1)$ coincide with x. Multiplication is always possible in the set $\mathscr{L}(X,x)$ of all loops with base point x. We denote by $\pi(X,x)$ the quotient of $\mathscr{L}(X,x)$ with respect to the equivalence relation $\sim$, and call it the *fundamental* (or the *Poincaré*) *group* of X at x. That this is a group via the product on classes of loops follows from §2. The identity element is the class of the constant loop. The inverse of the class of a loop γ is the class of the loop γ^{-1}. We often denote this group by $\pi_1(X,x)$ and call it the *first homotopy group* of X.

Remark 2. Let $\hat{X}$ be the arcwise connected component of the point x in X. We obviously have $\mathcal{L}(X,x) = \mathcal{L}(\hat{X},x)$ and the equivalence relation is the same, hence $\pi(X,x) = \pi(\hat{X},x)$. The fundamental group is of interest to us only when X is arcwise connected.

THEOREM 1. *If* X *is arcwise connected, then* $\pi(X,x) \cong \pi(X,y)$ *for all* $x, y \in X$.

Let γ be a path joining x to y. Define a mapping $\Gamma: \pi(X,y) \to \pi(X,x)$ by associating with each loop ℓ at y the loop ℓ' at x defined by $\ell' = \gamma \cdot \ell \cdot \gamma^{-1}$. The mapping Γ is a group homomorphism because $\gamma \cdot \ell \cdot \ell_1 \cdot \gamma^{-1} \sim \gamma \cdot \ell \cdot \gamma^{-1} \cdot \gamma \cdot \ell_1 \cdot \gamma^{-1}$. We can show that the mapping Γ': $\pi(X,x) \to \pi(X,y)$, defined with the aid of the path γ^{-1}, is the inverse of Γ. That follows from the equivalences $\ell \sim \gamma^{-1} \cdot \gamma \cdot \ell \cdot \gamma^{-1} \cdot \gamma$ and $\ell' \sim \gamma \cdot \gamma^{-1} \cdot \ell' \cdot \gamma \cdot \gamma^{-1}$. Therefore, Γ is an isomorphism.

Remark 3. By Theorem 1, the group $\pi(X,x)$ "does not depend" on the base point x chosen. We often write $\pi_1(X)$ without other specification. However, the above isomorphism Γ does depend on the particular path γ chosen and there does not exist a canonical isomorphism between $\pi(X,x)$ and $\pi(X,y)$. This is why we are usually careful to indicate the base point chosen.

A space X is said to be *simply connected* if $\pi(X,x)$ contains only the identity element, i.e., $\pi(X,x) = \{1\}$ (this definition does not depend on x). The space X consisting of a single point x is an example of a simply connected space (there is only one loop).

Effect of a continuous mapping. Let f: X → Y be a continuous mapping and let $y = f(x)$. By associating with each loop $\gamma \in \mathscr{L}(X,x)$ the loop $f \circ \gamma \in \mathscr{L}(Y,y)$, we induce a mapping on the equivalence classes:

$$\pi(f): \pi(X,x) \to \pi(Y,y)$$

which is a *group homomorphism.* [We write in general f: $(X,x) \to (Y,y)$ if $y = f(x)$.]

PROPOSITION 7. *If* f: $(X,x) \to (Y,y)$ *and* g: $(Y,y) \to (Z,z)$, *then*

$$\pi(g \circ f) = \pi(g) \circ \pi(f).$$

Furthermore,

$$\pi(id(X)) = id(\pi(X,x)).$$

The proof is easy.

COROLLARY. *Any two homeomorphic spaces have isomorphic fundamental groups.*

Let f: $(X,x) \to (Y,y)$ be a homeomorphism and let $g = f^{-1}$. It follows from Proposition 7 that

$$\pi(f) \circ \pi(g) = \pi(\mathrm{id}(Y)) = \mathrm{id}$$

and that

$$\pi(g) \circ \pi(f) = \mathrm{id},$$

which proves that $\pi(f)$ is an isomorphism.

Homotopic mappings (with base point). Let (X,x) and (Y,y) be two spaces with base point. Two continuous mappings:

$$f_0, f_1: (X,x) \to (Y,y)$$

are said to be *homotopic* (relative to the base points), denoted by $f_0 \sim f_1$, if there exists a continuous mapping F: X x I → Y such that:

$$F(\{x\} \times I) = y,$$

$$F(X \times \{0\}) = f_0, \text{ and } F(X \times \{1\}) = f_1.$$

THEOREM 2. *Homotopic mappings* f_0, f_1: $(X,x) \to (Y,y)$ *induce the same homomorphism*

$$\pi(f_0) = \pi(f_1)\colon \pi(X,x) \to \pi(Y,y).$$

In fact, if ℓ is a loop of X at x, then the loops $f_0 \circ \ell$ and $f_1 \circ \ell$ are homotopic. A homotopy between these loops is defined by $(t,s) \mapsto F(\ell(t),s)$ where F is a homotopy between f_0 and f_1.

A *homotopy equivalence* is a continuous mapping $f\colon (X,x) \to (Y,y)$ which possesses an "inverse to within homotopy," that is, a continuous mapping $g\colon (Y,y) \to (X,x)$ such that

$$f \circ g \sim \mathrm{id}(Y), \text{ and } g \circ f \sim \mathrm{id}(X).$$

In this case the pointed spaces (X,x) and (Y,y) are said to be *homeotopic* (or *homotopy-equivalent*).

COROLLARY. *Homeotopic spaces have isomorphic fundamental groups.*

In fact, we have

$$\pi(f \circ g) = \pi(\mathrm{id}(Y)) = \mathrm{id} \text{ and } \pi(g \circ f) = \pi(\mathrm{id}(X)) = \mathrm{id}.$$

Example 2. A space that is homeotopic to a point (also called a *contractible* space) is simply connected. This is the case for all convex sets of R^n.

Example 3. Every plane annulus

$$C = \{r_1 < |z| < r_2\}$$

is homotopy-equivalent to a circle. If we take as the circle, the circle S of radius $\dfrac{r_1 + r_2}{2}$, then the injection $S \to C$ admits as inverse to within a homotopy, the radial projection onto S.

Example 4. The plane with two points removed is homotopy-equivalent to the figure formed by the digit 8. The plane with n points removed is homeotopic to the boundary of an n-leafed rose.

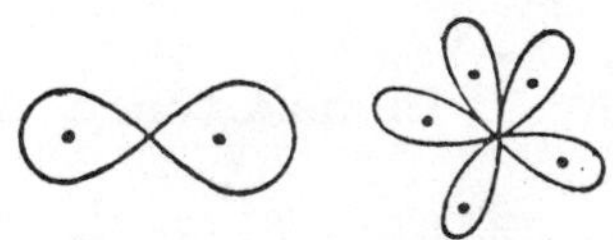

Example 5. The Möbius strip is homotopy-equivalent to a circle.

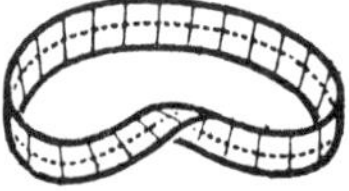

§4. BASIC EXAMPLES: CIRCLE, TORUS

A) *The fundamental group of the circle*

Let S^1 be the circle having equation $x^2 + y^2 = 1$ and let a be the point $(1,0)$ of S^1.

THEOREM 3. *The fundamental group of the circle is isomorphic to* Z.

We are going to construct a homomorphism $d\colon \pi(S^1,a) \to Z$ and show that it is an isomorphism. To define d, we first show that every class of $\pi(S^1,a)$ contains a differentiable loop and then define d for the differentiable loops.

Denote by $r\colon R^2\backslash\{(0,0)\} \to R^2$ the radial projection with respect to the origin onto the circle S^1. This is given by the formula

$$r(x,y) = \left(\frac{x}{\sqrt{x^2 + y^2}}, \frac{y}{\sqrt{x^2 + y^2}}\right).$$

Note that r is a differentiable mapping.

LEMMA 1. *Two loops f_0 and f_1 of (S^1,a) such that $d(f_0,f_1) = \sup_{t \in I} |f_0(t) - f_1(t)| \leq k < 1$ are homotopic.*

For all $s \in [0,1]$, we have $sf_1(t) + (1 - s)f_0(t) \neq (0,0)$, and the deformation $f_s(t) = r((1 - s)f_0(t) + sf_1(t))$ determines the homotopy.

LEMMA 2. *Every loop f of S^1 is homotopic to a differentiable loop ϕ that is close to f.*

A differentiable loop ϕ is a loop ϕ: $I \to S^1$ which, considered as a mapping in the plane, is differentiable. According to the Weierstrass-Stone theorem,[1] by approximating the two coordinates of f: $I \to R^2$ by a polynomial, a differentiable mapping ϕ_0: $I \to R^2$ is obtained such that $|f(t) - \phi_0(t)| \leqq \frac{\varepsilon}{4}$. If necessary, we can add a linear function (everywhere less than $\frac{\varepsilon}{4}$), so that $\phi_0(0) = \phi_0(1) = a$ and $|f(t) - \phi_0(t)| \leqq \frac{\varepsilon}{2}$. The loop $\phi(t) = r(\phi_0(t))$ is differentiable. It is homotopic to f by the deformation $r(s\,\phi_0(t) + (1 - s)f(t))$. In fact, it suffices to assume that $\varepsilon < 2$ so that $s\phi_0(t) + (1 - s)f(t)$ is never at (0,0). Moreover,

$$|\phi(t) - f(t)| \leqq |\phi(t) - \phi_0(t)| + |\phi_0(t) - f(t)| \leqq |2\,\phi_0(t) - f(t)| \leqq \varepsilon,$$

which proves that ϕ is as close to f as we wish.

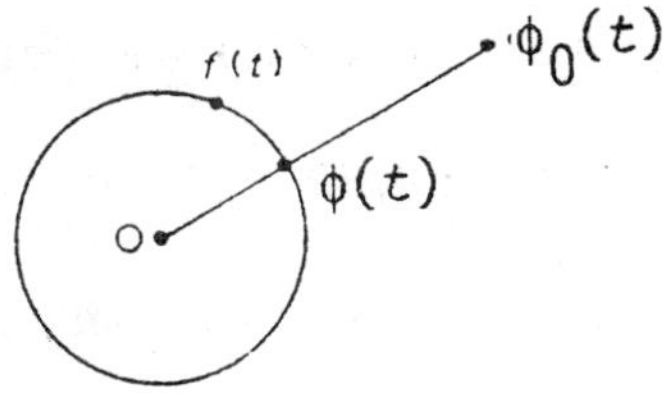

Degree of a differentiable loop. By definition, the degree of a differentiable loop ϕ is $d(\phi) = \frac{1}{2i\pi}\int_0^1 \frac{d\phi}{\phi}$. Set $\theta(\phi(t)) = \frac{1}{i}\int_0^t \frac{d\phi}{\phi}$; by comparing the derivatives of both sides of this equation, we see that

---

[1]Other theorems from analysis are good for this use. For example, we can form the convolution of f with a differentiable bell-shaped function.

$\phi(t) = e^{i\theta(\phi(t))}$. The function $\theta(\phi(t))$, which is continuous in t, is the argument of the exponential expression for $\phi(t)$ and is zero for $t = 0$. Therefore, the degree $d(\phi) = \frac{\theta(\phi(1))}{2\pi}$ is an integer, which intuitively represents the number of windings actually traversed by $\phi(t)$.

LEMMA 3. *Two close differentiable loops have the same degree.*

Let ϕ_0 and ϕ_1 be two differentiable loops such that $|\phi_0(t) - \phi_1(t)| \leqq \varepsilon < 1$. We set

$$\phi_s(t) = r((1 - s)\phi_0(t) + s\phi_1(t)).$$

Hence, both the differential $d\phi_s$ and the degree $d(\phi_s) = \frac{1}{2i\pi}\int_0^1 \frac{d\phi_s}{\phi_s}$ depend continuously on s. The function $d(\phi_s)$ is continuous in s and assumes only integer values, hence is constant (by the intermediate value theorem).

If f is a continuous loop, its *degree*, $d(f)$, is defined to be the common degree of all the differentiable loops that are close to f. This degree depends continuously on f because two close loops are close to the same differentiable loop. Hence, $d(f)$ depends only on the homotopy class of f (using the argument in Lemma 3).

a) *The mapping* $d\colon \pi(S^1, a) \to Z$ *is a homomorphism.* In fact, for differentiable loops, we have

$$d(\phi\cdot\phi') = \int_0^{\frac{1}{2}} \frac{d\phi(2\theta)}{\phi(2\theta)} + \int_{\frac{1}{2}}^1 \frac{d\phi'(2\theta - 1)}{\phi'(2\theta - 1)}$$

$$= \int_0^1 \frac{d\phi(t)}{\phi(t)} + \int_0^1 \frac{d\phi'(t)}{\phi'(t)} = d(\phi) + d(\phi').$$

b) *The homomorphism d is surjective.* If $n \in Z$, the loop $t \mapsto e^{2i\pi nt}$ has degree n.

c) *The homomorphism d is injective.* Let ϕ be differentiable such that $d(\phi) = \frac{\theta(\phi(1))}{2\pi} = 0$. The deformation $\phi_s(t) = e^{is\theta(\phi(t))}$ is a homotopy between ϕ and the constant loop: since $\theta(\phi(1)) = 0$, we have that, for all s, ϕ_s is really a loop.

This completes the proof of Theorem 3.

Remark 4. The group $\pi_1(S^1)$ is abelian. This is always the case for the fundamental group of a topological group.

B) *Application: The Brouwer fixed point theorem*

Let D^2 be the unit disk in the plane.

THEOREM 4. *Every continuous mapping $f: D^2 \to D^2$ has a fixed point.*

Assume the contrary, that for every $x \in D^2$, $x \neq f(x)$. Then x and $f(x)$ determine a line that intersects the boundary S^1 of D^2. Let $\phi(x)$ be the intersection point that is closer to x than to $f(x)$. The mapping $\phi: D^2 \to S^1$ is continuous and $\phi|S^1 = id(S^1)$. After Proposition 8, we show

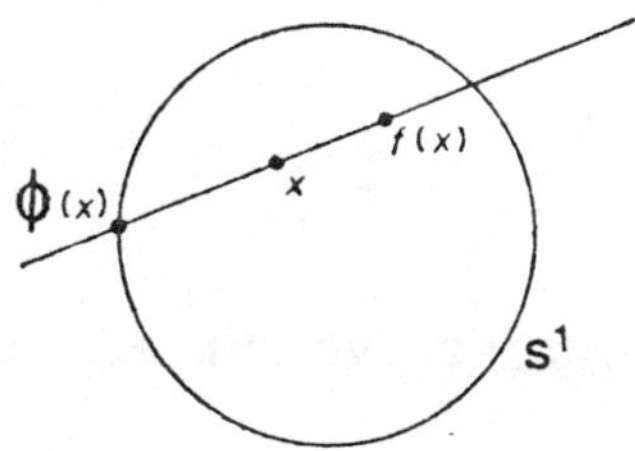

that this is impossible.

Let A be a subset of X and let $x \in A$. A mapping $r: X \to A$ is a *retraction* if $r|A = id(A)$.

PROPOSITION 8. *Let $r: X \to A$ be a retraction and let $i: A \to X$ be the canonical injection; the homomorphism*

$$\pi(i)\colon \pi(A,x) \to \pi(X,x)$$

has $\pi(r)$ as its left inverse (and hence $\pi(i)$ is injective).

In fact, $r|A = \mathrm{id}(A)$ means that $r \circ i = \mathrm{id}(A)$ and hence $\pi(r) \circ \pi(i) = \mathrm{id}(\pi(A,x))$.

We now conclude the proof of Theorem 4. Since $\pi_1(S^1) = Z$ (by Theorem 3) and $\pi_1(D^2) = \{1\}$ (by Example 2), there can be no retraction of D^2 onto S^1; otherwise (by Proposition 8) we would have an injection from Z into {1}.

C) *The fundamental group of the torus*

THEOREM 5. *The fundamental group of the torus is Z x Z.*

Let $T = S^1 \times S^1$ be the torus. Theorem 5 follows from a more general theorem. Let (X,x) and (Y,y) be two pointed spaces and let p_X: $(X \times Y,(x,y)) \to (X,x)$ and p_Y: $(X \times Y,(x,y)) \to (Y,y)$ be the obvious projections. The homomorphisms $\pi(p_X)\colon \pi(X \times Y,(x,y)) \to \pi(X,x)$ and $\pi(p_Y)\colon \pi(X \times Y,(x,y)) \to \pi(Y,y)$ define a homomorphism $p\colon \pi(X \times Y,(x,y)) \to \pi(X,x) \times \pi(Y,y)$.

THEOREM 6. *The homomorphism*

$$p\colon \pi(X \times Y,(x,y)) \to \pi(X,x) \times \pi(Y,y)$$

is an isomorphism.

The homomorphism p is surjective (i.e., onto) since selecting a loop at x in X and a loop at y in Y determines a loop at (x,y) in X x Y, whose projections are the selected loops. It is injective (i.e., one to one) since the homotopies in X and in Y are the projections of a homotopy in $(X \times Y,(x,y))$.

CHAPTER II

THE VAN KAMPEN THEOREM

If a space X is the union of two arcwise connected open sets U_1 and U_2 having arcwise connected intersection $U_1 \cap U_2$, the Van Kampen theorem (also known as the Seifert-Van Kampen theorem) enables us to calculate the fundamental group of X from those of U_1, U_2 and $U_1 \cap U_2$. In practice, by a judicious choice of U_1 and U_2, the Van Kampen theorem enables us to calculate $\pi_1(X)$ for many spaces X.

§1. THE FUNDAMENTAL GROUP OF THE SPHERE

THEOREM 1. *The sphere is simply connected.*

Let $S^2 \subset R^3$ be the sphere having equation $x^2 + y^2 + z^2 = 1$ and let $a = (1,0,0)$.

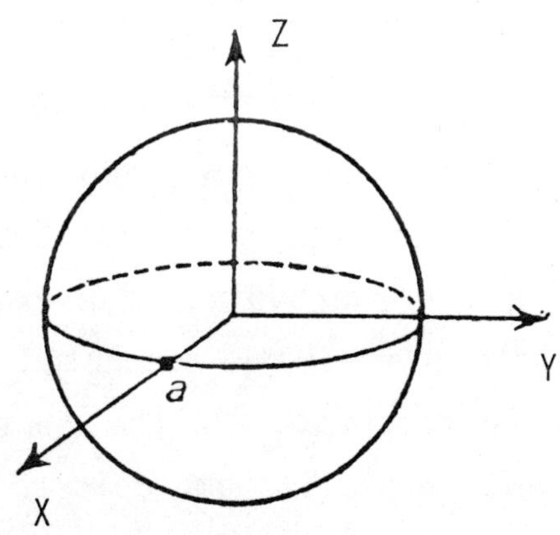

Let U_1 be the set of points of S^2 having z-coordinate greater than -1/2 and let U_2 be the set of points of S^2 having z-coordinate less than 1/2. The sphere is the union of U_1 and U_2. By using the stereographic projection, we can see that both U_1 and U_2 are homeomorphic to an open plane disk (hence are arcwise connected and simply connected) while $U_1 \cap U_2$ is homeomorphic to an annulus (hence is arcwise connected). Then Theorem 1 follows from the special case of the Van Kampen theorem proved below.[1]

PROPOSITION 1. *Suppose* $X = U_1 \cup U_2$, *where* U_1 *and* U_2 *are arcwise connected, simply connected open sets having arcwise connected intersection* $U_1 \cap U_2$. *Then* X *is simply connected.*

Proof of Proposition 1

LEMMA 1. *Let* $f\colon Y \to X$ *be a continuous mapping and let* $\{U_j\}_{j \in J}$ *be an opening covering of* X. *Then* $\{f^{-1}(U_j)\}_{j \in J}$ *is an open covering of* Y.

This is obvious.

LEMMA 2 (the existence of a Lebesgue number). *Let* $\{V_j\}_{j \in J}$ *be an open covering of a compact metric space* Y. *There is a number* $\varepsilon > 0$ *such that each ball of radius* $r \leqq \varepsilon$ *is contained in one of the open sets of the covering.*[2]

---

[1]Here we give a sketch of a direct proof for Theorem 1. Consider a loop f based at a in S^2 and assume that its image $f(I)$ does not contain all the points of the sphere and avoids, for example, a point $x \in S^2$. In the space $S^2\setminus\{x\}$ which is simply connected (it is homeomorphic to the plane), the loop is homotopic to the constant map c_a at a. It remains to show that every loop g is homotopic to a loop f which misses a point. If h is a *polygonal* loop close to g in $R^3\setminus\{0\}$, the radial projection f of h on S^2 has the required property.

[2]The number ε is called a *Lebesgue number* for the covering $\{V_j\}_{j\in J}$.

Each point $y \in Y$ is contained in some open set $V_{j(y)}$ and hence there is a number $\alpha(y) > 0$ such that the ball $B(y,\alpha(y))$ with center y and radius $\alpha(y)$ is contained in $V_{j(y)}$. We shall show that the infimum ε of the $\alpha(y)$ is positive. Assume the contrary. Then in particular, for each n there is a point y_n such that the ball $B(y_n,1/n)$ is not contained in any V_j. Since the metric space Y is compact, the sequence $\{y_n\}$ has[1] an accumulation point y. Let N be a natural number such that $d(y_N,y) < \frac{1}{2}\alpha(y)$ and $1/N < \frac{1}{2}\alpha(y)$. The ball $B(y_N,1/N)$ is contained in $V_{j(y)}$, which is a contradiction.

Now, let $x \in U_1 \cap U_2$ and let $f\colon I \to X$ be a loop at x. We wish to show that f is homotopic to the constant map c_x at x. To this end, it suffices to show that f is homotopic to a product $\phi_1 \cdot \phi_2 \cdot \ldots \cdot \phi_n$, where each ϕ_k is a loop at x and is contained in U_1 or U_2. Since U_1 and U_2 are simply connected, each ϕ_k is homotopic to c_x and their product is also homotopic to c_x.

The open sets $f^{-1}(U_1)$ and $f^{-1}(U_2)$ cover $I = [0,1]$. For n sufficiently large (Lemma 2), each of the segments $[\frac{k-1}{n}, \frac{k}{n}]$, where $1 \leqq k \leqq n$, is contained in one or the other of these two open sets. For each $k = 1,\ldots,n-1$, we choose a path γ_k which joins x to $x_k = f(\frac{k}{h})$. Since U_1, U_2 and $U_1 \cap U_2$ are arcwise connected, we may assume that γ_k is in U_1, U_2 or $U_1 \cap U_2$ depending on whether x_k is in U_1, U_2 or $U_1 \cap U_2$. Denote

---

[1]A necessary and sufficient condition for a metric space to be *compact* is that each sequence has an accumulation point, i.e., has a convergent subsequence.

by f_k the path which is the restriction of f to $[\frac{k-1}{n}, \frac{k}{n}]$ and set

$$\phi_1 = f_1 \cdot \gamma_1^{-1},$$

$$\phi_k = \gamma_{k-1} \cdot f_k \cdot \gamma_k^{-1} \text{ if } 2 \leqq k \leqq n-1,$$

$$\phi_n = \gamma_{n-1} \cdot f_n.$$

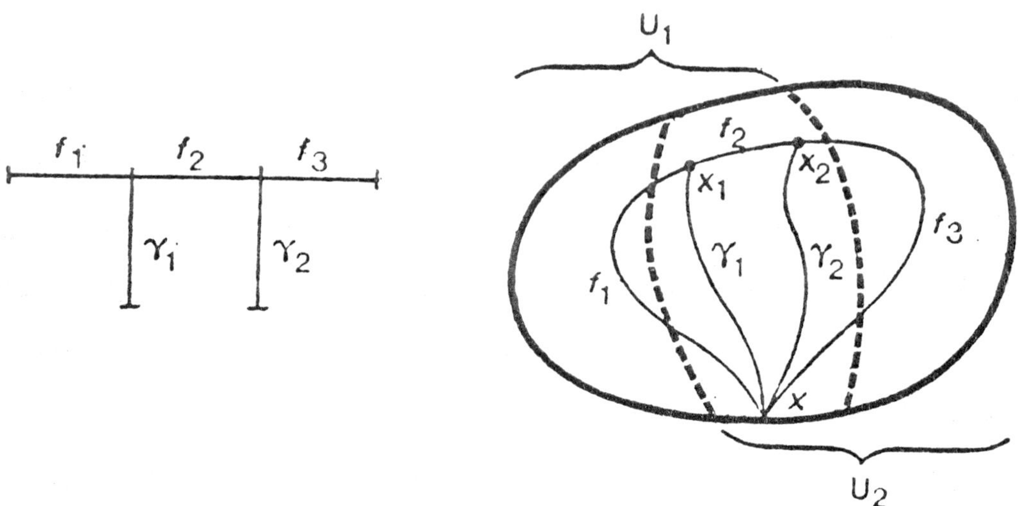

Since $[\frac{k-1}{n}, \frac{k}{n}] \subset f^{-1}(U_j)$, then ϕ_k is a loop at x in U_j. On the other hand,

$$f = f_1 \cdot f_2 \cdot \ldots \cdot f_n \sim \phi_1 \cdot \phi_2 \cdot \ldots \cdot \phi_n,$$

which completes the proof.

Remark 1. Proposition 1 enables us to prove, more generally, that the sphere $S^n \subset R^{n+1}$, determined by the equation $x_1^2 + \ldots + x_{n+1}^2 = 1$, is simply connected for $n \geqq 1$.

Example 1. The complement of the origin 0 in R^3 (which is homotopy-equivalent to S^2) is simply connected.

§2. RUDIMENTS OF GROUP THEORY[1]

Free (non-abelian) groups on two generators

Let E be the set of all the finite words (the empty word included) that can be formed by juxtaposing the symbols a^p and b^q, p and $q \in Z$. Given a word, the following reductions are permissible:

(i) Replace an occurrence of two consecutive symbols a^p, a^q by the symbol a^{p+q}, and similarly replace an occurrence of two consecutive symbols b^m, b^n by the symbol b^{m+n}.

(ii) Delete a^0 and b^0.

A word in which no further reductions are possible is called a *reduced* word. A reduced word consists of a sequence of symbols that are alternately of the form a^p and b^q, where the exponents are not zero. We can verify without difficulty that each word reduces to exactly one reduced word. Let Z*Z (or F(a,b)) denote the set of reduced words with the following multiplication: the product $m \cdot m'$ of two words is the reduced word corresponding to the (not necessarily reduced) word obtained by writing m followed by m'. With this operation Z*Z is a group. The empty word is the neutral element and $(a^{-p_n} b^{-q_n} \dots a^{-p_1} b^{-q_1})$ is the inverse of $(b^{q_1} a^{p_1} \dots b^{q_n} a^{p_n})$, etc.

The mappings $p \mapsto (a^p)$ and $q \mapsto (b^q)$ give two monomorphisms $\tilde{a}$, $\tilde{b}$: $Z \to Z*Z$. Furthermore, if G is a group, and if ϕ and ψ: $Z \to G$ are

---

[1]The groups that we are concerned with are not necessarily commutative. In general we adopt multiplicative notation, the neutral element being denoted by 1. In particular, the additive group Z will be represented as $\{a^n | n \text{ integer}\}$, with multiplication law defined by $a^n \cdot a^{n'} = a^{n+n'}$.

morphisms (homomorphisms), then there is a unique morphism $\theta: Z*Z \to G$ such that $\theta \circ \tilde{a} = \phi$, $\theta \circ \tilde{b} = \psi$. This morphism is determined by $\theta((a^p)) = \phi(p)$, $\theta((b^q)) = \psi(q)$.

Free group on a set S

One may generalize the preceding notion. Let S be a set. Instead of forming words using only the two letters *a* and *b*, use all the elements $s \in S$. In particular, if S is a finite set with *n* elements, we obtain a free group on *n* generators. We denote this group by F(S).

The Free Product of Two Groups (also called *the sum of two groups*)

This is another generalization inspired by the construction of Z*Z. Let G_1 and G_2 be two groups. Consider the set of finite words constructed using the elements of G_1 and the elements of G_2. It is allowable to replace two consecutive letters g_1, g_1' if they are both in the group G_1 by the unique letter $g_1'' = g_1 \cdot g_1' \in G_1$, and similarly with elements from G_2. In addition, we delete the neutral elements. Analogous to the previous construction, a reduced word is a finite sequence of non-neutral elements coming alternately from G_1 and from G_2. The set of reduced words with the obvious multiplication determines a group G_1*G_2, the *free product* of G_1 and G_2.

The mappings $g_1 \mapsto (g_1)$ and $g_2 \mapsto (g_2)$ give monomorphisms:

$$\theta_1: G_1 \to G_1*G_2 \text{ and } \theta_2: G_2 \to G_1*G_2.$$

We frequently denote the element $(g_1) \in G_1*G_2$ by g_1. Furthermore, if G is a group and if $\mu_1: G_1 \to G$ and $\mu_2: G_2 \to G$ are morphisms, then there is a unique morphism $\mu: G_1*G_2 \to G$ such that $\mu \circ \theta_1 = \mu_1$, and $\mu \circ \theta_2 = \mu_2$. That is, there is a unique morphism μ such that the diagram

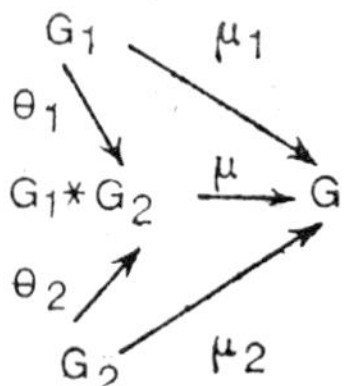

is commutative.

Remark 2. Such a "universal" property suffices to characterize the free product $G_1 * G_2$ up to isomorphism.

Remark 3. If G_2 is the trivial group $\{1\}$, then the morphism $\theta_1: G_1 \to G_1 * \{1\}$ is an isomorphism.

The Free Product (of two groups) with Amalgamation

Assume that G_0, G_1, G_2 are groups and that $\phi_1: G_0 \to G_1$ and $\phi_2: G_0 \to G_2$ are homomorphisms. Let N be the smallest normal[1] subgroup of $G_1 * G_2$ that contains all the elements of the form $(\phi_1(g), \phi_2(g)^{-1})$ and $(\phi_2(g), \phi_1(g)^{-1})$, where $g \in G_0$, and let $\mu: G_1 * G_2 \to (G_1 * G_2)/N$ be the natural map. We denote $(G_1 * G_2)/N$ by $G_1 *_{G_0} G_2$ and we say that it is the free product of G_1 and G_2 amalgamated on G_0 by ϕ_1 and ϕ_2 (or, when the context is clear, we call this simply the *amalgamated product*).

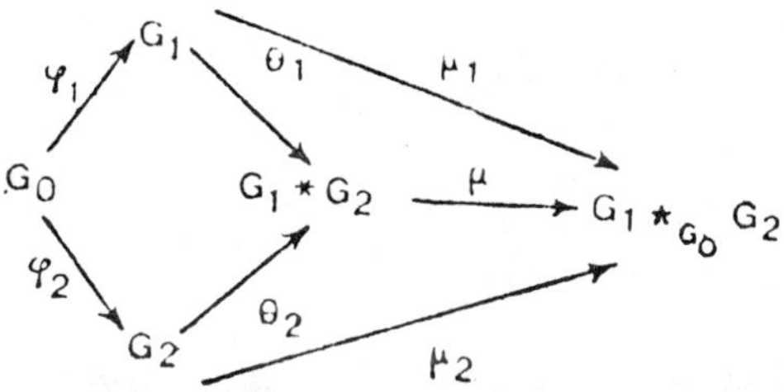

---

[1]We say that a subgroup N of H is *normal* (or *invariant*) if every inner automorphism $x \to h^{-1}xh$ of H sends N into N. This is equivalent to saying that the multiplication of H induces a (well-defined) multiplication on the set of cosets H/N. This is equivalent to saying that N is the kernel of a homomorphism of H into another group. If K is a subgroup of H, the intersection $\overrightarrow{K}$ of all the normal subgroups of H containing K consists of all the conjugates $h^{-1}kh$, $h \in H$, $k \in K$ and their products. It is the smallest normal subgroup of H containing K.

The morphisms $\mu_1 = \mu \circ \theta_1$ and $\mu_2 = \mu \circ \theta_2$ satisfy $\mu_1 \circ \phi_1 = \mu_2 \circ \phi_2$ since the elements $\theta_1 \circ \phi_1(g)$ and $[\theta_2 \circ \phi_2(g)]^{-1}$ have a product that is in the kernel of μ. Moreover, if H is a group and $\psi_1\colon G_1 \to H$ and $\psi_2\colon G_2 \to H$ are homomorphisms such that $\psi_1 \circ \phi_1 = \psi_2 \circ \phi_2$, then there exists a unique homomorphism $\psi\colon G_1 *_{G_0} G_2 \to H$ for which $\psi_1 = \psi \circ \mu_1$ and $\psi_2 = \psi \circ \mu_2$. In fact, there exists a unique homomorphism $\psi'\colon G_1 * G_2 \to H$ such that $\psi_1 = \psi' \circ \theta_1$ and $\psi_2 = \psi' \circ \theta_2$. But the property $\psi_1 \circ \phi_1 = \psi_2 \circ \phi_2$ shows that the kernel of ψ' contains all elements of the form $(\phi_1(g),\phi_2(g)^{-1})$, $g \in G_0$, and hence contains N. The homomorphism ψ' induces a homomorphism on the quotient by N.

Characterization of the elements of N

In order for a (not necessarily reduced) word to represent an element of $G_1 * G_2$ which is in N, it is necessary and sufficient that one can reduce it to the neutral element (the empty word) by a sequence of operations of the following two types:

A) Replace $\phi_1(g)$ by $\phi_2(g)$, $g \in G_0$, and conversely.

B) Replace two consecutive letters g_i, g_i' (where g_i and $g_i' \in G_i$, $i = 1,2$) by the letter $g_i'' = g_i \cdot g_i' \in G_i$, and, conversely, decompose a letter g_i'' into g_i, g_i' if $g_i'' = g_i \cdot g_i'$. (This operation does not change the corresponding element of $G_1 * G_2$. It enables us to attain the reduced word.)

Every element of N is a product of elements of the form $(a^{-1},\phi_1(g),\phi_2(g)^{-1},a)$, $a \in G_1 * G_2$, $g \in G_0$, or of their inverses. By an operation A and an operation B we reduce it to (a^{-1},a) which reduces to ( ) in $G_1 * G_2$.

Conversely, we shall show that if $(m_1,\phi_1(g),m_2)$ is in N, then $(m_1,\phi_2(g),m_2)$ is also in N. First note that $(\phi_1(g),m_2,m_1) =$

$(m_1^{-1})(m_1,\phi_1(g),m_2)(m_1)$ is again in the normal subgroup N. Multiplying by $(\phi_2(g),\phi_1(g)^{-1}) \in$ N, we obtain, after a reduction B, $(\phi_2(g),m_2,m_1)$. Now another conjugation yields $(m_1,\phi_2(g),m_2)$ which is thus in N.

Remark 4. If $G_0 = \{1\}$, then N is the trivial group and the amalgam is isomorphic to $G_1 * G_2$.

Remark 5. If $G_2 = \{1\}$, then N is the normal subgroup generated by the $\phi_1(g)$, $g \in G_0$. Furthermore, $G_1 * G_2$ is isomorphic to G_1 (cf. Remark 2). The amalgam is the quotient of G_1 by the smallest normal subgroup, $\overline{\phi_1(G_0)}$, containing $\phi_1(G_0)$.

§3. THE VAN KAMPEN THEOREM

THEOREM 2. *Assume that* $X = U_1 \cup U_2$, *where* U_1 *and* U_2 *and also* $U_0 = U_1 \cap U_2$ *are arcwise connected open sets. Let* $x \in U_0$. *Then* $\pi(X,x)$ *is the free product of* $\pi(U_1,x)$ *and* $\pi(U_2,x)$ *amalgamated on* $\pi(U_0,x)$:

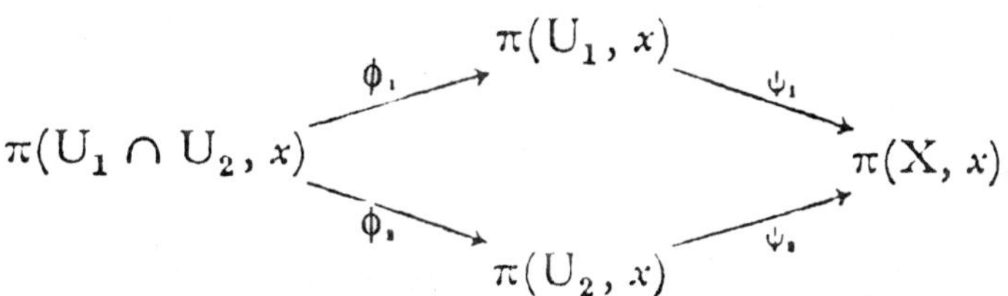

The morphisms ϕ_1, ϕ_2, ψ_1, ψ_2 are induced by the obvious inclusion mappings. We clearly have $\psi_1 \circ \phi_1 = \psi_2 \circ \phi_2$. The theorem asserts that the morphism

$$\psi' : \pi(U_1) * \pi(U_2) \to \pi(X)$$

associated (see §2) with ψ_1 and ψ_2 is an isomorphism. To this end, we shall show that the homomorphism $\psi' : \pi(U_1) * \pi(U_2) \to \pi(X)$ is surjective and that N contains the kernel of ψ'.

Part 1 (onto). Let f be a loop in X based at x. We must show that f is homotopic to a product of loops at x each of whose images is contained in the space U_1 or U_2. This was done in §1 as part of the proof of Proposition 1.

Part 2 (the kernel of ψ' is contained in N). For this we are first going to give a geometric interpretation and then prove it.

*Part 2 (interpretation).** Let $h_1, h_2, \ldots, h_n$ be loops at x in the spaces U_1 and U_2. They determine an element $(\overline{h_1}, \overline{h_2}, \ldots, \overline{h_n})$ of $\pi(U_1,x) * \pi(U_2,x)$ whose image under ψ' is represented by the loop $h_1 \cdot h_2 \cdot \ldots \cdot h_n$ in X at x. If this loop is homotopic to c_x in X we will show that in $\pi(U_1) *_{\pi(U_0)} \pi(U_2)$ we can transform the symbol $(\overline{h_1}, \overline{h_2}, \ldots, \overline{h_n})$ into $(\overline{c_x})$. This will be done using the following operations:

H. If h_k is a loop homotopic to h_k' with respect to x, then we can replace $\overline{h_k}$ by $\overline{h_k'}$.

A'. If h_k is a loop in U_1 whose image is in $U_1 \cap U_2$, then we may consider it as a loop in U_2, and conversely.

B'. If h_k is a loop in U_i and $h_k = h_k' \cdot h_k''$ where h_k', h_k'' are loops in U_i then we can replace $\overline{h_k}$ by $\overline{h_k'}$, $\overline{h_k''}$.

This would suffice to prove Theorem 2 since if $(\overline{h_1}, \overline{h_2}, \ldots, \overline{h_n})$ is in the kernel of ψ', then $h_1 \cdot h_2 \cdot \ldots \cdot h_n \sim c_x$ and then we would get that in the amalgam $(\overline{h_1}, \overline{h_2}, \ldots, \overline{h_n}) = (\overline{c_x})$. The operations A' and B' are the geometric interpretations of the operations A and B, and H corresponds to

---

*Throughout this proof we are using $\overline{h}$ to designate the homotopy class of h with respect to x.

the fact that homotopic loops define the same element of the fundamental group. We also note that the operation A' passes over imperceptibly to a geometric line of reasoning because we identify a loop in U_1 whose image is also in U_2 with the loop in U_0 defined by the "same" mapping.

Part 2 (geometric proof). Let $(\bar{h}_1,\bar{h}_2,\ldots,\bar{h}_n)$ be in the kernel of ψ', and let $H\colon I \times I \to X$ be a homotopy in X between $h_1 \cdot h_2 \cdot \ldots \cdot h_n$ and the constant loop c_x. The open sets $H^{-1}(U_1)$ and $H^{-1}(U_2)$ cover the compact metric space I x I. Let $\varepsilon > 0$ be a Lebesgue number (see §1, Lemma 2) of this covering, and let p be a sufficiently large integer so that $\frac{\sqrt{2}}{pn} \leq \varepsilon$. The image under H of each square $[\frac{k-1}{np}, \frac{k}{np}] \times [\frac{j-1}{np}, \frac{j}{np}]$, $1 \leq k \leq np$, $1 \leq j \leq np$, is entirely contained in one of the open sets U_1 or U_2, denoted by $U_{i(k,j)}$. *We limit ourselves to the case $n = 1$ and $p = 2$.* It is easy to see how in a similar manner one could establish the general case, but the complexity of the notation required would render the proof difficult to follow.

The following figure represents the square I x I, the domain of H. This square is decomposed into four squares of edge length 1/2, numbered 1 to 4. The image under H of each of these squares is entirely contained in one of the open sets U_1 or U_2. The restrictions of H to each of the edges of the small squares are paths whose names we have noted on the figure.

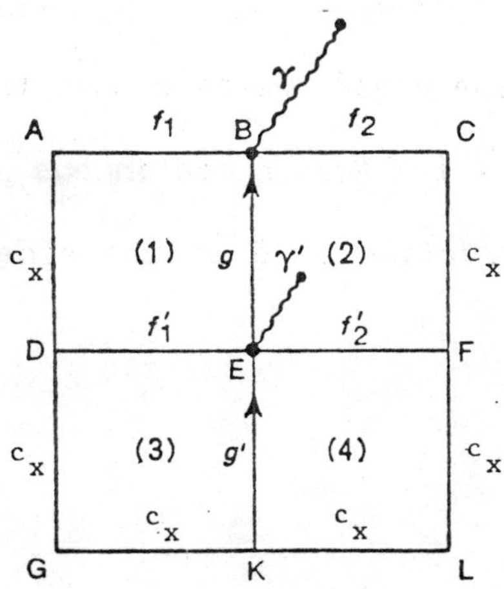

The word that we wish to reduce in the amalgamated product is represented by $h_1 = f_1 \cdot f_2$.

We have:

$(\overline{h_1}) = (\overline{f_1 \circ \gamma^{-1} \circ \gamma \circ f_2})$ using operation H in the open set containing h_1,

$= (\overline{f_1 \circ \gamma^{-1}}, \overline{\gamma \circ f_2})$ using operation B,

$= (\overline{c_x \circ f_1' \circ g \circ \gamma^{-1}}, \overline{\gamma \circ f_2})$ applying H to the open set U_1 which contains the image of the square (1),

$= (\overline{c_x \circ f_1' \circ \gamma'^{-1}}, \overline{\gamma' \circ g \circ \gamma^{-1}}, \overline{\gamma \circ f_2})$ same reason,

$= (\overline{c_x}, \overline{f_1' \circ \gamma'^{-1}}, \overline{\gamma' \circ g \circ \gamma^{-1}}, \overline{\gamma \circ f_2})$ by B',

$= (\overline{c_x}, \overline{c_x \circ c_x \circ g' \circ \gamma'^{-1}}, \overline{\gamma' \circ g \circ \gamma^{-1}}, \overline{\gamma \circ f_2})$ using H,

$= (\overline{g' \circ \gamma'^{-1}}, \overline{\gamma' \circ g \circ \gamma^{-1}}, \overline{\gamma \circ f_2})$ by B.

Similarly, working on $\gamma \circ f_2$, we obtain

$$(\overline{h_1}) = (\overline{g' \circ \gamma'^{-1}}, \overline{\gamma' \circ g \circ \gamma^{-1}}, \overline{\gamma \circ g^{-1} \circ \gamma'^{-1}}, \overline{\gamma' \circ g'^{-1}}),$$

and applying B' and H twice, we obtain $(\overline{h_1}) = (\overline{c_x})$, which concludes the proof.

In the course of the proof, we have used the following lemma.

LEMMA 3. *If* $\delta\colon I \times I \to X$ *is a continuous mapping,* γ_0 *designates the path which is the restriction of* δ *to the side* AB *of* $I \times I$, *and* γ_1 *the composition of the restrictions to the sides* AC, CD *and* DB, *then* γ_0 *and* γ_1 *are equivalent* (homotopic to the endpoints $\delta(A)$ and $\delta(B)$).

The homotopy is easily deduced from the schema of the figure below.

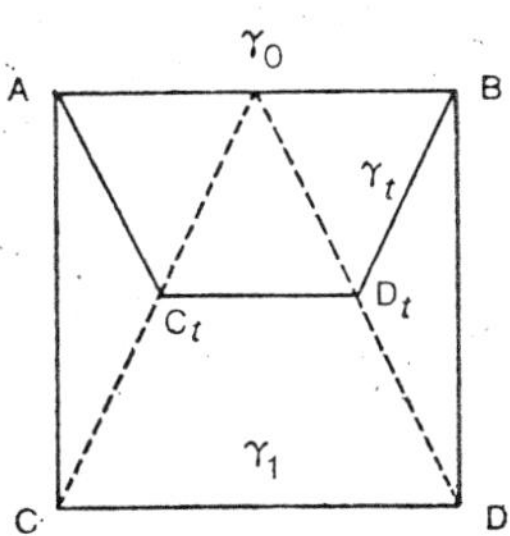

§4. BASIC APPLICATIONS

Remark 6. We will often apply Theorem 2 when $X = A_1 \cup A_2$, where A_1 and A_2 as well as $A_0 = A_1 \cap A_2$ are arcwise connected sets. And, furthermore, where A_1 and A_2 have open neighborhoods U_1 and U_2 for which the injections $A_0 \subset U_0 = U_1 \cap U_2$, $A_1 \subset U_1$ and $A_2 \subset U_2$ are homotopy-equivalent. If $x \in A_0$, the group $\pi(X,x)$ is the product of $\pi(A_1,x)$ and $\pi(A_2,x)$ amalgamated by $\pi(A_0,x)$.

PROPOSITION 2. *The fundamental group of the figure formed by the figure 8 is a free group on two generators.*

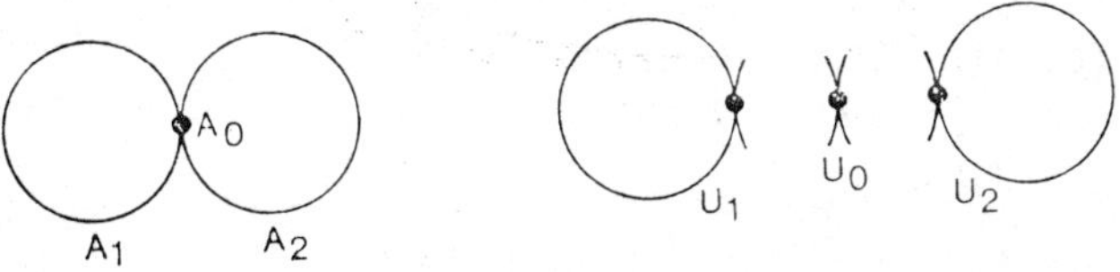

The above figure is the union of two circles A_1 and A_2 whose intersection reduces to a point A_0. We easily see that one has the conditions of Remark 6 and the fundamental group is a free group on two generators, one being a generator of $\pi(A_1)$ and the other a generator of $\pi(A_2)$:

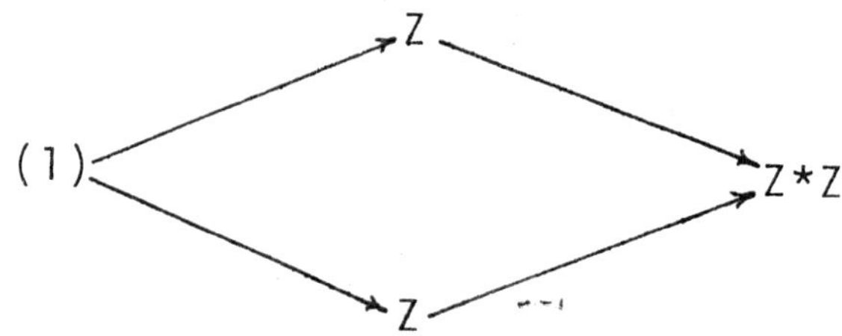

In general, we could prove by induction on n, that the fundamental group of the n-leafed rose (see Chapter I, Example 4) is a free group on n generators.

PROPOSITION 3. *The fundamental group of the punctured torus is a free group on 2 generators.*

Let $T_1 = S^1 \times S^1$ be the torus parametrized by the pairs (θ,ϕ) of numbers modulo 2π. The circle $(0,\phi)$, $0 \leqq \phi \leqq 2\pi$, is the meridian m (oriented by increasing ϕ) and the circle $(\theta,0)$, $0 \leqq \theta \leqq 2\pi$, the longitude ℓ (oriented by increasing θ). The punctured torus is the space V_1 which is the complement of a "square" $C = (0 < \theta < \varepsilon, 0 < \phi < \varepsilon)$[1].

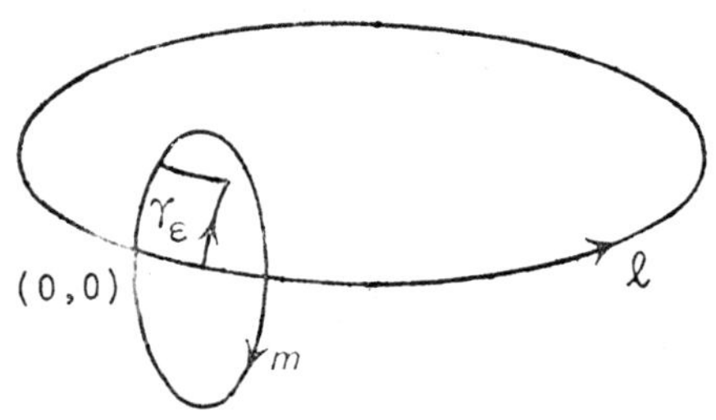

[1]We obtain a homeomorphic space by removing only one open disk contained in the square C. More generally, it follows from a theorem of L. ANTOINE (1920) that we still obtain a space homeomorphic to V_1 if we remove from the torus the interior of a closed set K which is the homeomorphic image of the disk D^2.

This space is the union of the cylinder $A_1 = (\varepsilon \leqq \theta \leqq 2\pi) = [\varepsilon,2\pi] \times m$ and the cylinder $A_2 = (\varepsilon \leqq \phi \leqq 2\pi) = \ell \times [\varepsilon,2\pi]$, whose intersection is the square $A_0 = [\varepsilon,2\pi] \times [\varepsilon,2\pi]$. As in Proposition 2, the group $\pi(V_1,(0,0))$ is a free group on two generators, the generator m of $\pi(A_1,(0,0))$ and the generator ℓ of $\pi(A_2,(0,0))$.

The boundary of V_1 is the boundary γ_ε of the square $[0,\varepsilon] \times [0,\varepsilon]$. Going around once, starting in the direction of ℓ, γ_ε is a loop at $(0,0)$ in V_1. The family of boundaries γ_t of the squares $[0,t] \times [0,t]$, $\varepsilon \leqq t \leqq 2\pi$ is a homotopy between γ_ε and the loop $\ell m \ell^{-1} m^{-1}$. Another application of Theorem 2, now to $T_1 = V_1 \cup C$, permits us to retrieve $\pi(T_1)$:

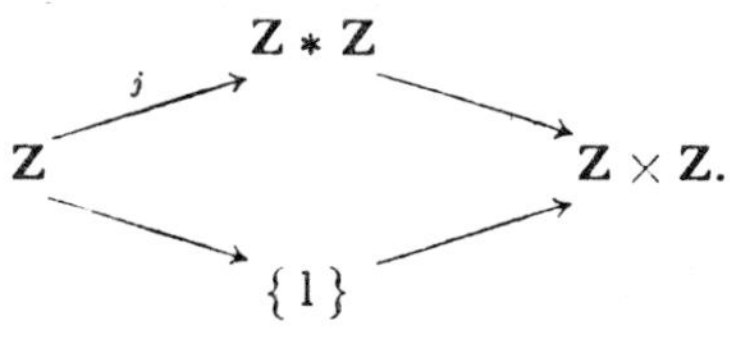

The homomorphism j sends the generator γ_ε of $\pi(V_1 \cap C)$ to the commutator $[\ell,m] = \ell m \ell^{-1} m^{-1}$. The group $\pi(T_1) = Z * Z/(\ell m \ell^{-1} m^{-1})$ (Remark 5) is the *free abelian group* on the two generators m and ℓ.

§5. GLUING OF SPACES

Let X and Y be two metric spaces, A a compact subset of X and f: A → Y an isometry. Let $\mathcal{R}$ be the equivalence relation on $X \cup Y$ which identifies a and $f(a)$, $a \in A$. The class $\bar{x}$ of $x \in X$ is $\{x, f(x)\}$ if $x \in A$, $\{x\}$ otherwise; the class $\bar{y}$ of $y \in Y$ is $\{y, f^{-1}(y)\}$ if $y \in f(A)$, $\{y\}$ otherwise. On the space $Z = (X \cup Y)/\mathcal{R}$ we can define a distance by:

$$d(\bar{x},\bar{x}') = d(x,x') \text{ if } x,\ x' \in X,$$

$$d(\bar{y},\bar{y}') = d(y,y') \text{ if } y,\ y' \in Y,$$

$$d(\bar{y},\bar{x}) = d(\bar{x},\bar{y}) = \inf_{a \in A} (d(x,a) + d(f(a),y)) \text{ if } x \in X,\ y \in Y.$$

In the last formula, the infimum of the continuous function $d(x,a) + d(f(a),y)$ is attained at a point a of the compact set A. We shall show that d is a metric. If $d(\bar{x},\bar{y}) = 0$, there exists a point $a \in A$ such that $d(x,a) = 0$ and $d(f(a),y) = 0$, whence $x = a$, $y = f(a)$ and $\bar{x} = \bar{y}$. Symmetry is obvious. The triangle inequality is proven by cases. We prove as an example the case where $x,\ z \in X$ and $y \in Y$. Then

$$d(\bar{x},\bar{y}) = d(x,a) + d(f(a),y), \text{ for some } a \in A,$$

$$d(\bar{y},\bar{z}) = d(y,f(a')) + d(a',z), \text{ for some } a' \in A,$$

$$d(\bar{x},\bar{z}) = d(x,z) \leqq d(x,a) + d(a,a') + d(a',z).$$

But:

$$d(a,a') = d(f(a),\ f(a')) \leqq d(f(a),y) + d(y,f(a'))$$

since f is an isometry; whence the result.

The canonical inclusions i: $X \to Z$ and j: $Y \to Z$ are isometries. In particular $Z\backslash j(Y) = i(X\backslash A)$ is open so that $i(X)$ and $j(Y)$ are closed in Z. The specification of a continuous mapping g: $Z \to T$ is equivalent to specifying the continuous mappings h: $X \to T$ and ℓ: $Y \to T$ such that $h|A = \ell \circ f$ and we have $h = g \circ i$ and $\ell = g \circ j$ (see Chapter I, Proposition 6). In addition, if X and Y are compact, then Z is compact.

We shall not give the general[1] definition of a space obtained by identifying x and $f(x)$, where f is a mapping from a subset A of X into X

---

[1]If $\mathcal{R}$ is an equivalence relation on a space X, we can equip the quotient space $X/\mathcal{R}$ with a natural topology: the largest topology that renders the natural map $X \to X/\mathcal{R}$ continuous. It is not true, in general, that this topology is metrizable if X is metrizable (see N. BOURBAKI, *General Topology*, Chapter IX, Addison-Wesley, 1973).

itself. In special cases, appropriate definitions are easy to obtain. For example, consider the sphere S^2 and the mapping f: $S^2 \to S^2$ which sends x to $-x$. The space obtained by identifying x and $f(x)$ is the *projective plane* $P_2(R)$, (the space of lines through the origin in R^3). If we equip P_2 with the metric defined by $d(\bar{x},\bar{y}) = \inf d(x,y)$, $x \in \bar{x}$, $y \in \bar{y}$, the canonical mapping p: $S^2 \to P_2$ is a locally isometric continuous mapping;[1] it preserves distances less than $\sqrt{2}$.

The space P_2 *is compact*, since the canonical mapping p has compact domain and P_2 as its image. More explicitly, let $\{\bar{x}_n\}$ be a sequence of points of P_2 and for each n let $x_n \in S^2$ be one of the two points whose image is $\bar{x}_n$ under the mapping p: $S^2 \to P_2$; since S^2 is a compact space, the sequence $\{x_n\}$ has an accumulation point x and $p(x)$ is an accumulation point of $\{\bar{x}_n\}$.

Here is a construction of P_2 by gluing. We observe that S^2 is obtained by gluing together two caps D_1 and D_2 to an equatorial annulus C. The restriction of p to D_1 or to D_2 is an isometry. The image of C under p is a Möbius strip M whose boundary is the isometric image of the upper boundary of C. The space P_2 is the union of M and the disk $D = p(D_1)$, the intersection of M and D being their common boundary which is homeomorphic to a circle.

---

[1]From this it follows that the mapping p: $S^2 \to P_2$ is a *covering projection*. The theory of covering projections (see W. MASSEY, *Algebraic Topology: An Introduction*, Springer, 1977) is intimately related to that of the fundamental group. In particular, if f: $Y \to X$ is a covering projection and if Y is connected and simply connected, then for $x \in X$ there is a bijection between $f^{-1}(x)$ and the fundamental group $\pi(X,x)$. We shall give a different proof that $\pi_1(P_2)$ is the group with two elements (§7, Proposition 7).

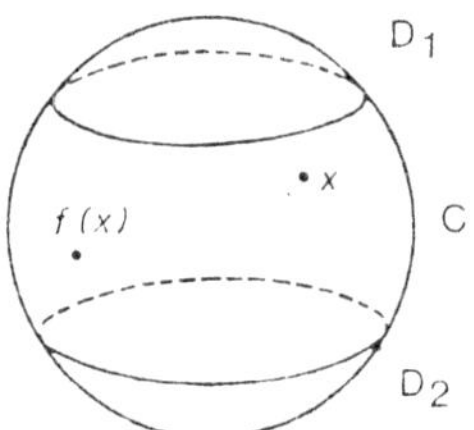

The Van Kampen theorem often allows us to calculate the fundamental group of a space obtained by gluing.

§6. COMPACT SURFACES

Denote by T_2 the space obtained by gluing together two copies V_1' and V_1'' of V_1 along their "circular" boundaries S' and S" by the homeomorphism $(S',0') \to (S'',0'')$ which reverses the orientation. We denote $S = S' = S''$ and say that T_2 is *the oriented surface of genus 2.*

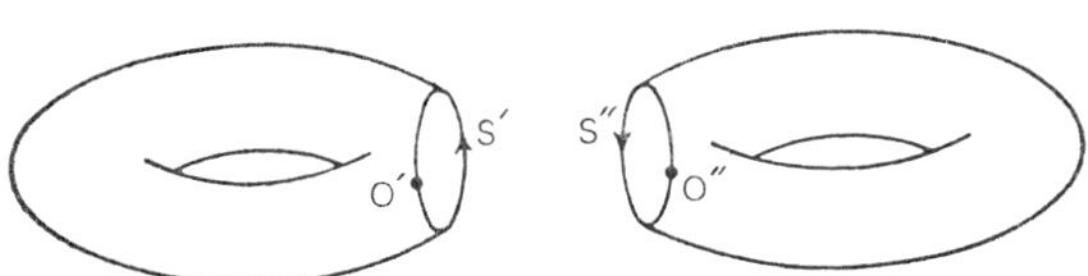

PROPOSITION 4. *The fundamental group of the oriented surface of genus two is the quotient of the free group on four generators* $F(a_1,b_1,a_2,b_2)$ *by the normal subgroup generated by* $[a_1,b_1][a_2b_2] = a_1b_1a_1^{-1}b_1^{-1}a_2b_2a_2^{-1}b_2^{-1}$.

We reapply Theorem 2 with $A_0 = S$, $A_1 = V_1'$ and $A_2 = V_1''$.

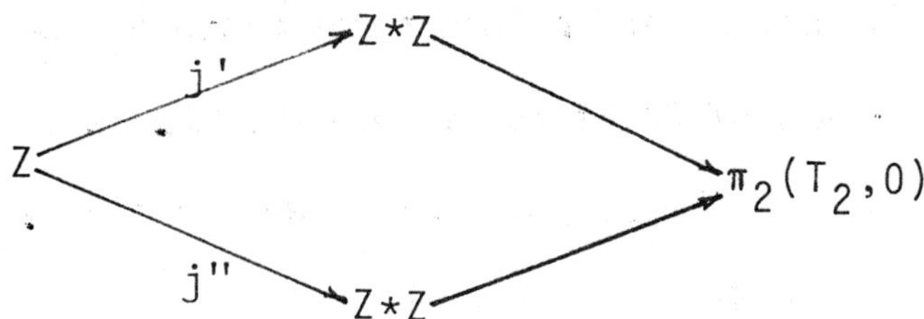

The image of the generator of S under j' is the class of S', that is, $[\ell,m]$ (see §4) and under j'' it is the inverse to the class of S", that is, $[\ell',m']^{-1} = m'\ell'm'^{-1}\ell'^{-1}$. Setting $a_1 = m$, $b_1 = \ell$, $a_2 = m'$, $b_2 = \ell'$, we see that $\pi(T_2,0)$ is the quotient of $Z*Z*Z*Z$ by the normal subgroup N corresponding to the relation $[a_1,b_1][a_2,b_2]^{-1}$ (that is, the normal subgroup generated by $[a_1,b_1][a_2,b_2]$).

We now denote by V_2 the space obtained by removing a small square from T_2.

PROPOSITION 5. *The fundamental group of* V_2 *is a free group on four generators.*

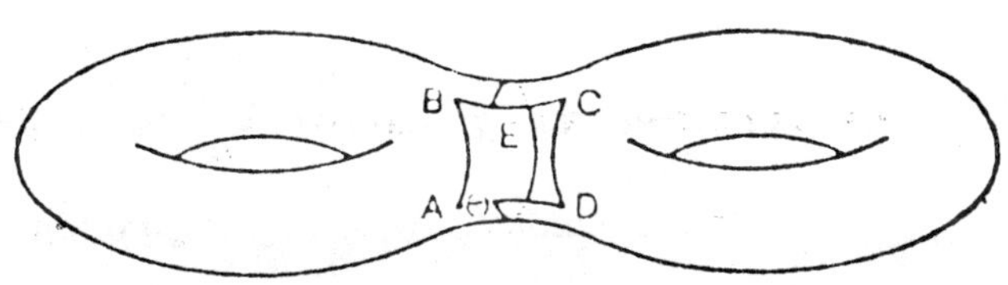

Assume the square C to be symmetric with respect to S (see figure), and let $I = S \cap V_2$; i.e., I is a space homeomorphic to a segment. We have $V_2 = A' \cup A''$, where A' and A" are homeomorphic to V_1 and $A' \cap A'' = I$. Theorem 2 yields $\pi_1(V_2,0)$ is a free group on four generators a_1,b_1,a_2,b_2 and the boundary γ of the square C represents the class $\bar{\gamma} = [a_1,b_1][a_2,b_2]$. In fact, the boundary γ = OABECDO of the square C is a loop at 0 in V_2 which is homotopic to the composition of the loops OABEO and OECDO (the journey EO and return takes place on the segment I). But we know that the class of OABEO is $[a_1,b_1]$ and that the class of OECDO is $[a_2,b_2]$; whence $\bar{\gamma} = [a_1,b_1][a_2,b_2]$. We obtain $\pi_1(T_2)$ by a new application of Theorem 2.

We will assume for V_2 as for V_1 that the above results do not depend on the place where we pierce the small hole, round or square (cf. p. 30, footnote 1). Then define, inductively, the oriented surface T_n of genus n, as well as V_n:

DEFINITION. For $n \geqq 1$, V_n is the space obtained by removing a small disk from T_n and T_{n+1} is obtained by gluing together V_n and V_1 along their boundaries. We define $T_0 = S^2$ and $V_0 = D^2$.

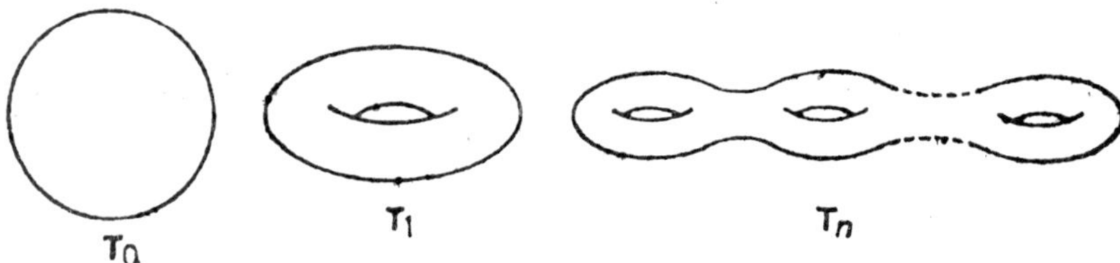

PROPOSITION 6. *The fundamental group of* V_n *is a free group on 2n generators* $a_1,b_1,\ldots,a_n,b_n$, *the boundary* γ *of* V_n *represents the class* $\bar{\gamma} = [a_1,b_1][a_2,b_2]\ldots[a_n,b_n]$. *The fundamental group of* T_n *is isomorphic to the quotient of the preceding free group by the normal subgroup generated by* $\bar{\gamma}$.

Proof, by induction, similar to that of Proposition 5.

Remark 7. Let V and V' be two manifolds of dimension n (see p. 2, footnote). Let D and D' be 'nice', small, closed disks of dimension n in open subsets of V, respectively V', that are homeomorphic to R^n. Denote by V # V' the space obtained by gluing together V\int(D) and V'\int(D') via a homeomorphism of their boundaries S and S'. These boundaries are homeomorphic to the sphere S^{n-1}. We can verify that V # V' is a manifold of dimension n, called the *connected sum* of V and V'. Thus, $T_{k+1} = T_k$ # T_1. If $n \geqq 3$, the sphere S^{n-1} is simply connected (§1, Remark 1), so that an application of Theorem 2 and Remarks 4 and 5 of §2 gives $\pi(V) = \pi(V\backslash int(D))*\pi(D) = \pi(V\backslash int(D))$. Using Theorem 2 again, we obtain $\pi(V \# V') = \pi(V)*\pi(V')$. This does not hold when $n = 2$ (cf. Propositions 5 and 6).

§7. NON-ORIENTABLE SURFACES[1]

The real projective plane is the union of a disk D and a Möbius strip M with intersection S = D $\cap$ M homeomorphic to a circle (§5). The fundamental group of M is infinite cyclic (see Chapter I, §3, Example 5) and its boundary S is a loop homotopic to twice the generator (in additive notation).

PROPOSITION 7. *The fundamental group of the projective plane is a group with two elements.*

This follows from Theorem 2 and the diagram:

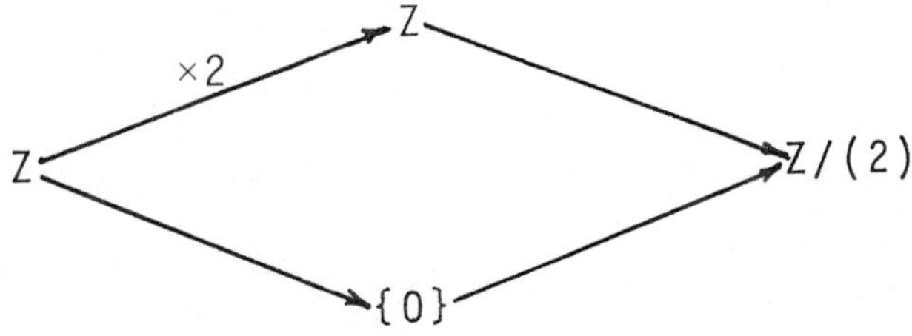

In multiplicative notation, $\pi(P_2)$ is the quotient of the free group on one generator α by the subgroup generated by α^2.

---

[1]The concept of orientation will be made precise later (see Chapter V, §4).

The connected sum of two copies of P_2 is the *Klein bottle* $U_2 = P_2 \# P_2$ which is obtained by gluing together two Möbius strips along their boundaries (which are homeomorphic to a circle). By applying Theorem 2, one proves that $\pi(U_2) = Z*Z/(\alpha^2\beta^2)$, where α and β are the two generators. In general, we set $U_{n+1} = U_n \# P_2$, and call U_n the *non-oriented surface of genus n*. By induction, we obtain the following:

PROPOSITION 8. *The fundamental group of the non-oriented surface* U_n *of genus n is the quotient of the free group on n generators* $\alpha, \beta, \ldots, \nu$ *by the normal subgroup generated by* $\alpha^2\beta^2\ldots\nu^2$.

Remark 8. We can prove that $T_1 \# P_2$ is homeomorphic to $U_2 \# P_2$, and then by induction, that U_{2n+1} is homeomorphic to $T_n \# P_2$ and U_{2n+2} to $T_n \# U_2$ (see Chapter V).

CHAPTER III

DIFFERENTIABLE FUNCTIONS AND MANIFOLDS

It is convenient to review the definition and some properties of differentiable functions before considering a particularly interesting class of metric spaces--the differentiable manifolds. A differentiable manifold is a subspace of a Euclidean space R^n which is sufficiently "nice" so that we can talk about differentiable functions on the manifold.[1] We are primarily interested here in manifolds of dimension 2, i.e., *surfaces*. The reader can find the proofs of the results indicated in this chapter and related developments in J. DIEUDONNÉ, *Foundations of modern analysis* (Academic Press, 1978) and H. CARTAN, *Calcul différentiel et formes différentielles* (Hermann, 1967).

§1. DIFFERENTIABLE MAPPINGS

Definition. Let U be an open subset of R^p *and let* f be a continuous mapping of U into R^q. We say that f is a *differentiable mapping* at $a \in U$ if there exists a linear mapping $Df(a)\colon R^p \to R^q$ such that

---

[1]We can give an intrinsic definition of a structure of a differentiable manifold on a metric space without assuming that it is a subset of R^n (see N. BOURBAKI, *Variétés différentielles et analytiques* (Résultats, §5); but it can be proved that this definition is no more general than the one we give.

$$\frac{\|f(x) - f(a) - [Df(a)]((x - a))\|}{\|x - a\|} \to 0$$

when $x \to a$ $(x \neq a)$.

The linear mapping $Df(a) \in L(R^p, R^q)$ possessing this property is the *differential* of f at a. The differential of f at a, if it exists, is unique. The rank of the linear mapping $Df(a)$ (that is, the dimension of the image) is called the *rank* of f at a.

A linear mapping ϕ: $R \to R$ is determined by the number $\phi(1)$. The usual derivative $f'(a)$ of f: $R \to R$ is precisely equal to $[Df(a)](1)$. If f: $R^p \to R^q$ is a linear mapping, it is differentiable and furthermore, for every $a \in R^p$, $Df(a) = f \in L(R^p, R^q)$.

Mappings of class C^r. If f is differentiable at every point $a \in U$, the mapping Df: $U \to L(R^p, R^q)$ which associates $Df(x)$ with $x \in U$ is the *differential* of f. If Df is continuous, f is said to be of class C^1, or continuously differentiable. If Df is differentiable, its derivative $D(Df)(a) \in L(R^p, L(R^p, R^q))$ determines a bilinear mapping $D^2f(a)$: $R^p \times R^p \to R^q$. If Df is of class C^1, we say that f is of class C^2 and (by Schwarz's theorem) $D^2f(a)$ is symmetric bilinear. By induction, we say in general that f is of the class C^r if its differential Df is of class C^{r-1}, and that f is of class C^∞ if for all r the differentials D^rf are continuously differentiable.

The differential of a composite function (the chain rule). Let U be an open subset of R^p, V an open subset of R^q and f: $U \to R^q$, g: $V \to R^n$ be mappings. If f is differentiable at $a \in U$ and g is differentiable at $f(a) \in V$, then $g \circ f$ is differentiable at a and its differential is $D(g \circ f)(a) = Dg(f(a)) \circ Df(a)$: $R^p \to R^n$.

Partial derivatives. To define a mapping $f\colon U \to R^q$ is equivalent to specifying q functions $f_1,\dots,f_q$ of p variables $x_1,\dots,x_p$. The partial derivative $\frac{\partial f_i}{\partial x_j}$ at the point $a = (a_1,\dots,a_p)$ is the derivative at $a_j \in R$ of the composite mapping:

$$x_j \mapsto (a_1,\dots,x_j,\dots,a_p) \mapsto f_i(a_1,\dots,x_j,\dots,a_p).$$

It follows from the above that if f is differentiable at a, it possesses partial derivatives. The matrix representing $Df \in L(R^p,R^q)$ with respect to the canonical bases is the *Jacobian matrix*:

$$J(f) = \begin{pmatrix} \frac{\partial f_1}{\partial x_1} \cdots & \frac{\partial f_1}{\partial x_j} \cdots \\ \vdots & \vdots \\ \frac{\partial f_i}{\partial x_1} \cdots & \frac{\partial f_i}{\partial x_j} \cdots \\ \vdots & \vdots \end{pmatrix}.$$

The converse is false. However, a necessary and sufficient condition for f to be of class C^1 in U is that it has continuous partial derivatives in U.

§2. INVERSE FUNCTION THEOREM

Let $f\colon U \to R^q$ be a mapping of class C^r. If we assume that f is a bijection from U onto an open set $V = f(U)$ and if its inverse f^{-1} is differentiable, then we say that f is a C^r-*diffeomorphism* from U onto V. In that case we have $p = q$, $Df^{-1}(f(a)) = (Df(a))^{-1}$ and f^{-1} is of class C^r.

THEOREM 1 (Inverse Function Theorem). *Let* $f\colon U \to R^p$ *be a mapping of class* C^r. *If the differential is of rank* p *at some point* $a \in U$ (i.e., is an invertible linear mapping), *then there is an open neighborhood* A *of* a *such that the restriction* $f|A$ *is a* C^r*-diffeomorphism from* A *onto* $f(A)$.

Thus, the differential of a differentiable mapping provides information on the local properties of the latter.

COROLLARY (Implicit Function Theorem). *Let* U *be an open subset of* $R^p \times R^q$ *and let* $f\colon U \to R^q$ *be a mapping of class* C^r. *If at the point* $(a,b) \in U$ *the "partial differential"* $D_2f(a,b)\colon R^q \to R^q$ *has rank* q, *then there is a neighborhood* V *of* (a,b), *a neighborhood* W *of* a *in* R^p *and a mapping* $g\colon W \to R^q$ *of class* C^r *such that the following conditions are equivalent:*

i) $(x,y) \in V$ *and* $f(x,y) = f(a,b)$, *and*

ii) $x \in W$ *and* $y = g(x)$.

Moreover, the mapping $h\colon V \to R^p \times R^q$, *defined by* $h(x,y) = (x,f(x,y))$ *is a* C^r*-diffeomorphism from* V *onto* $h(V)$.

The partial differential $D_2f(a,b)$ is the differential of the mapping $y \to f(a,y)$ at the point $y = b \in R^q$. The second assertion of the corollary follows from Theorem 1 since the differential $Dh(a,b)\colon R^p \times R^q \to R^p \times R^q$ is determined by the block matrix

$$Dh(a,b) = \begin{pmatrix} \mathrm{id}(R^p) & 0 \\ D_1f(a,b) & D_2f(a,b) \end{pmatrix}.$$

Hence we see that $Dh(a,b)$ is a mapping of rank $p + q$, and therefore h is locally invertible. The inverse of h can be written as $h^{-1}(x,z) = (x,g(x,z))$ for some function g. It is easy to show that g properly restricted is "really only a function of x" and is the desired function for the first assertion of the corollary.

§3. DIFFERENTIABLE MANIFOLDS

Definition. A subset M of R^n is a class C^r *differentiable manifold* of dimension p if it has the following local property:

(P) *For each point* $a \in$ M *there is an open subset* U *of* R^p, *a neighborhood* W *of* a *in* R^n *and a mapping* ϕ: U $\to R^n$ *of class* C^r *having rank* p, *which is a homeomorphism from* U *onto* W $\cap$ M.

We say that ϕ is a *parametrization* of M in a neighborhood of a. After making a translation of U (if necessary), it may be assumed that $\phi(0) = a$; this is called a parametrization centered at a.

Property (P) is equivalent to the following:

(P') *For each point* $a \in$ M *there exist an open set* X *of* R^n (which contains the origin 0) *and a* C^r*-diffeomorphism* ψ *from* X *onto a neighborhood* W *of* a *in* R^n (which contains the origin 0) *such that* ψ(X $\cap$ R^p) = W $\cap$ M, *where* R^p *is the subspace* $(y_{p+1} = \cdots = y_n = 0)$ *of* R^n.

It is clear that (P') implies (P) if we set U = X $\cap$ R^p and $\phi = \psi|$U. The converse can be deduced from the following lemma.

LEMMA. *If* ϕ: U $\to R^n$ *is a parametrization of* M *centered at* a, *there is a neighborhood* X *of* 0 *in* R^n *and a* C^r*-diffeomorphism* ψ *from* X *onto a neighborhood* W' *of* a *in* R^n *such that* X $\cap$ R^p = U' *is contained in* U, *that* ψ(U') = W' $\cap$ M, *and that* $\psi|$U' $= \phi|$U' *is a parametrization of* M *centered at* a.

The mapping ϕ has rank p at 0 and we may assume, after a permutation of the coordinates x_i of R^n, that the determinant of the matrix $\left(\frac{\partial\phi_i}{\partial y_j}\right)$, $1 \leq i \leq p$, $1 \leq j \leq p$ is not zero. We see that the mapping ψ: U x $R^{n-p} \to R^n$, defined by

$$\psi_i(y_1,\ldots,y_n) = \phi_i(y_1,\ldots,y_p) \qquad 1 \leq i \leq p,$$

$$\psi_{p+k}(y_1,\ldots,y_n) = y_{p+k} + \phi_{p+k}(y_1,\ldots,y_p), \; 1 \leq k \leq n - p,$$

has rank n at 0 because its Jacobian matrix is

$$J(\psi) = \left(\begin{array}{c:c} \dfrac{\partial\phi_i}{\partial y_j} & \dfrac{\partial\phi_{p+k}}{\partial y_j} \\ \hdashline 0 & \text{id} \end{array}\right).$$

For X we take a sufficiently small neighborhood of $0 \in R^n$ so that $\psi|X$ is a diffeomorphism from X onto a neighborhood W' of a contained in W.

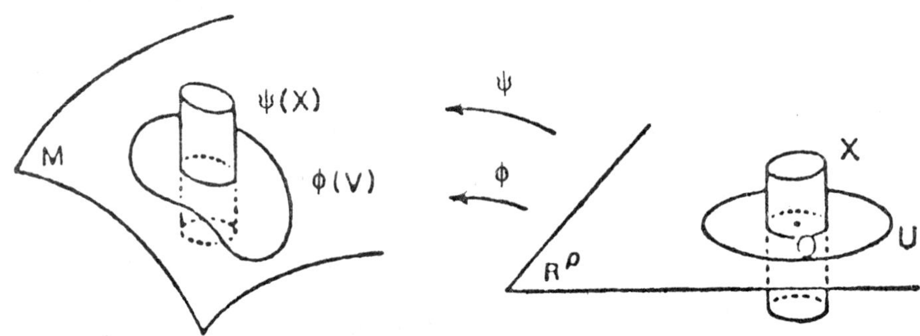

Remark 1. Since ψ is injective, it follows from the lemma that if U is replaced by a sufficiently small neighborhood of 0, we may assume that the parametrization ϕ could be extended to an injective map of a neighborhood in R^n. The mapping ϕ^{-1} is called a *system of local coordinates* in a neighborhood of a or a *chart* centered at a.

Remark 2. A differentiable manifold of dimension p is a manifold of dimension p in the sense of Footnote 1 on page 2 (Chapter I) since a parametrization is a homeomorphism.

Example 1. A manifold of dimension 1 is a *curve* in R^n and a manifold of dimension 2 is a *surface*.

Example 2. One can verify that the sphere

$$x^2 + y^2 + z^2 - 1 = 0$$

is a manifold of dimension 2 in R^3, as is the torus with parametric equations:

$$x = (3 + \cos \phi)\cos \theta,$$
$$y = (3 + \cos \phi)\sin \theta,$$
$$z = \sin \phi.$$

Change of parametrization. If ϕ: U → R^n is a parametrization of M in a neighborhood of a, U' an open subset of R^p and θ: U' → U is a C^r-diffeomorphism, then the mapping $\phi \circ \theta$: U' → R^n is a parametrization of M in a neighborhood of a. Conversely, let ϕ': U' → R^n be a parametrization of M; we shall show that $\phi^{-1} \circ \phi' = \theta$ is a C^r-diffeomorphism from U' onto U (we assume that ϕ and ϕ' are injective). The mapping θ: U' → U is bijective; it suffices to show that it has rank p at every point $y \in$ U'. We may assume that ϕ is centered at $\phi'(y)$. Let ψ: X → R^n be associated with ϕ as in the lemma and let V be a neighborhood of y in U' such that ϕ'(V) ⊂ ψ(X). The mapping $\psi^{-1} \circ \phi'$: V → R^n has rank p since ψ is a diffeomorphism and its image is contained in R^p; it coincides with $\phi^{-1} \circ \phi$ which therefore has rank p. Thus, every change of parametrization of a p-manifold is induced by a C^r-diffeomorphism from an open subset of R^p.

Differentiable mappings of a manifold into a manifold. Let f: M → M' be a continuous mapping. We say that f is *differentiable at* $a \in$ M if, for some parametrization ϕ of M centered at a, the mapping $f \circ \phi$: U → M' ⊂ R^n is differentiable at 0. This does not depend on the choice of ϕ (see above). We say that f is *differentiable* if it is differentiable at every point of M. The *rank* of f at a is the rank (independent of ϕ) of $f \circ \phi$ at 0. If f is bijective and if f^{-1} is differentiable, we say that f is a *diffeomorphism*. This is equivalent to saying that f is bijective and has rank equal to the dimension of M. In particular, if ϕ: U → M is a parametrization, then ϕ^{-1}: ϕ(U) → U is a diffeomorphism.

Tangent plane (for surfaces). Let ϕ: U → R^n be a parametrization of M centered at a. The image of the linear affine mapping of rank 2, $a + D\phi$: $R^2 \to R^n$, is a plane passing through a which is called the *tangent plane* $T_a(M)$. It is independent of the parametrization ϕ. Every other parametrization can be written as $\phi \circ \theta$ where θ is a diffeomorphism, and the image of $D(\phi \circ \theta) = D\phi \circ D\theta$ is that of $D\phi$. Let $x = x(t)$, $y = y(t)$ be a differentiable curve in R^2, passing through 0 for $t = 0$. The tangent vector at a to the curve $\phi(x(t), y(t))$, traced on the surface M, is the vector having origin a and endpoint $a + D\phi(x'(0), y'(0))$; it is in the tangent plane $T_a(M)$. We denote by $\frac{\partial}{\partial x}$ the tangent vector to $\phi(t,0)$ and by $\frac{\partial}{\partial y}$ the tangent vector to $\phi(0,t)$. These vectors generate $T_a(M)$ and every tangent vector A can be written uniquely as $\alpha\frac{\partial}{\partial x} + \beta\frac{\partial}{\partial y}$, which is the velocity vector of $\phi(\alpha t,\beta t)$. If $x = x(u,v)$, $y = y(u,v)$ is a local change of coordinates with differential $D\theta = \begin{pmatrix} x'_u & x'_v \\ y'_u & y'_v \end{pmatrix}$, then we have

$$\frac{\partial}{\partial u} = x'_u \frac{\partial}{\partial x} + y'_u \frac{\partial}{\partial y}, \text{ and}$$

$$\frac{\partial}{\partial v} = x'_v \frac{\partial}{\partial x} + y'_v \frac{\partial}{\partial y},$$

whence

$$A = \alpha \frac{\partial}{\partial x} + \beta \frac{\partial}{\partial y} = \alpha' \frac{\partial}{\partial u} + \beta' \frac{\partial}{\partial v}$$

with $(\alpha,\beta) = D\theta(\alpha',\beta')$.

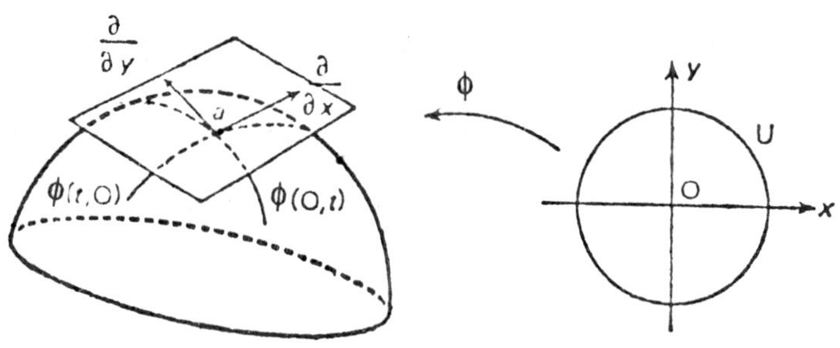

Tangent mapping. Let f: M → M' be a differentiable mapping and let ϕ be a parametrization of M in a neighborhood of a. The mapping which associates with the tangent vector $A = \alpha\frac{\partial}{\partial x} + \beta\frac{\partial}{\partial y}$ the vector having origin $b = f(a)$ and end point $b + D(f \circ \phi)(\alpha,\beta)$ is a mapping of the plane $T_a(M)$ into the plane $T_b(M')$. It does not depend on the parametrization chosen since

$$D(f\circ\phi\circ\theta)(\alpha',\beta') = D(f\circ\phi)\circ D(\theta)(\alpha',\beta') = D(f\circ\phi)(\alpha,\beta).$$

This mapping, denoted by Df, is called the *tangent mapping* to f at a.

§4. REGULAR VALUES OF A DIFFERENTIABLE FUNCTION

Let f: M → R be a differentiable function on a manifold of dimension p. A point a ∈ M is a *critical point* of f if $Df(a)$: $T_a(M) \to$ R is zero. We say that $f(a)$ is a *critical value* of f. A *regular value* is a number which is not a critical value (for any critical point a ∈ M).

PROPOSITION 1. *If y is a regular value of f: M → R, then $f^{-1}(y)$ = N is a manifold of dimension $p - 1$.*

To prove this, we assume that M is a surface.[1] We assume that N is not empty and let a ∈ N. If we find a neighborhood V of a in M and a system of local coordinates $\psi^{-1}(x) = (f(x),v(x))$ in V, one of them being the function f, then N is a manifold (a curve). It is parametrized by $\psi(y,v)$ since N is the set of points such that $f(x) = y$.

Let $\phi(u,v)$ be a parametrization of M centered at a. Since $Df(a) \neq 0$, we may assume (after perhaps interchanging u and v) that $Df\cdot\frac{\partial}{\partial u} = \frac{d}{du}(f \circ \phi(u,0))$ is not zero for $u = 0$. It then follows from the corollary

---

[1]Proposition 1 is nonetheless valid for every dimension p.

to Theorem 1 that $\theta(u,v) = (f \circ \phi(u,v),v)$ has rank 2 at 0 and that $(f(x),v(x)) = \theta \circ \theta^{-1}(x)$ is a system of local coordinates on M in a neighborhood of a.

Example 3. The value 0 is regular for a function $f(x,y,z)$ on R^3 if $f(x,y,z) = 0$ implies that at least one of the partial derivatives f'_x, f'_y, f'_z is not zero. In this case the equation $f(x,y,z) = 0$ determines a surface (cf. Example 2).

Example 4. Consider a surface M in R^3, the projection to the z-axis is a function--call it the *height function z*. Its derivative D_z is the projection of the tangent plane. This vanishes at critical points: these are the points of M having a horizontal tangent plane.

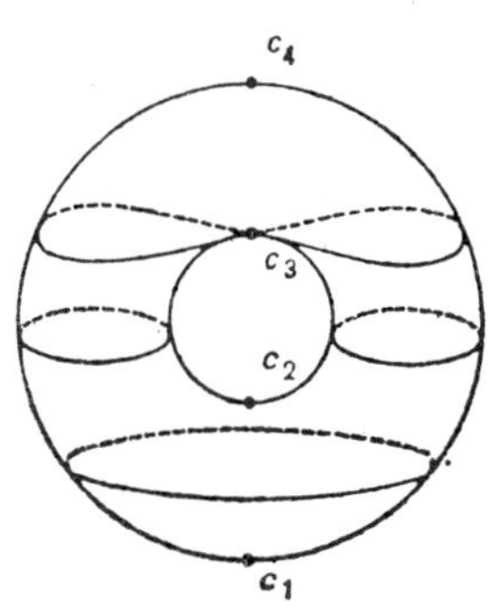

The figure shows a torus about the x-axis and four critical points of the height function. The critical values are the coordinates of these horizontal sections which are not curves (1-manifolds).

§5. CRITICAL POINTS

Let M be a surface, $f: M \to R$ a function of class C^r $(r \geqq 3)$ and $c \in M$ a critical point of f, that is $Df(c) = 0$. Let $\phi(x,y)$ be a parametrization of M centered at c; the function $g(x,y) = f(\phi(x,y))$ has a critical point at 0. The second derivative at 0 of g is an element $D(D(g))(0)$ of $L(R^2, L(R^2,R))$ with which there is canonically associated a symmetric bilinear mapping $D^2g \in \mathrm{Bil}(R^2 \times R^2, R)$.

This mapping is represented by the *Hessian matrix*,

$$H(g) = \begin{pmatrix} g''_{xx} & g''_{xy} \\ g''_{yx} & g''_{yy} \end{pmatrix} .$$

Let θ be a change of parametrization $x = x(u,v)$, $y = y(u,v)$. After the change of parametrization θ, we have $h(u,v) = g(\theta(u,v))$, whence

$$\frac{\partial h}{\partial u} = \frac{\partial}{\partial u}(g(\theta(u,v))) = g'_x x'_u + g'_y y'_u,$$

$$\frac{\partial^2 h}{\partial u^2} = g''_{xx}(x'_u)^2 + 2g''_{xy}x'_u y'_u + g''_{yy}(y'_u)^2 + g'_x x''_{uu} + g'_y y''_{uu},$$

and similarly for the other second partial derivatives. From this it follows that

$$D^2h(A'',B'') = D^2g(D\theta.A'',D\theta.B'') + Dg.D^2\theta(A'',B''),$$

where A'' and $B'' \in R^2$.

When 0 is a critical point of g, we have that $Dg = 0$, whence

(1) $$D^2h(A'',B'') = D^2g(D\theta.A'',D\theta.B'')$$

where

$$H(h) = {}^tJ(\theta).H(g).J(\theta).$$

In this case, we define a symmetric bilinear form on the tangent plane $T_c(M)$ by setting

$$D^2f(A,B) = D^2g(A',B'),$$

where $A = D\phi(A')$ and $B = D\phi(B')$. By equation (1), this does not depend on the choice of parametrization.

Remark 3. It is essential that c be a critical point so that only the (first order) differential of the change of parametrization θ appears in formula (1). In general, we cannot speak of the second order differential on a manifold without specifying the local coordinates, whereas the differential (§3) is always defined.

The characteristics of the bilinear form D^2f at a critical point c, and of the associated quadratic form are independent of the parametrization. We say that the critical point is *nondegenerate* if the determinant of the Hessian is not zero, i.e., if D^2f is nondegenerate. The *index* of the nondegenerate critical point c is the index of the quadratic form associated with D^2f. The index is the dimension of the largest subspace where the quadratic form is negative definite. The index is 0, 1 or 2 according as D^2f is *positive, hyperbolic*, or *negative*.

THEOREM 2 (Morse 1932).[1] *Let c be a nondegenerate critical point of f: $M \to R$, of class C^r $(r \geqq 3)$. There exists a parametrization ϕ with local coordinates (X,Y) in a neighborhood $\phi(U)$ of c such that*

$$f(\phi(X,Y)) = f(c) + \begin{cases} X^2 + Y^2 & \text{if the index is } 0 \\ X^2 - Y^2 & \text{if the index is } 1 \\ -X^2 - Y^2 & \text{if the index is } 2. \end{cases}$$

Let ψ be a parametrization of M centered at c with local coordinates (x,y) and let $g(x,y) = f(\psi(x,y))$. It suffices to show the assertion of the following lemma.

LEMMA. *Let g: $U \to R$ be a function of class C^r $(r \geqq 3)$ defined in a neighborhood U of 0 in R^2, for which 0 is a nondegenerate critical point of index i. Then there is a diffeomorphism $(x,y) = \theta(X,Y)$ of a neighborhood V of 0 in R^2 onto a neighborhood $\theta(V)$ of 0 in U such that:*

$$g(\theta)(X,Y)) = g(0) + \begin{cases} X^2 + Y^2 & \text{if } i = 0 \\ X^2 - Y^2 & \text{if } i = 1 \\ -X^2 - Y^2 & \text{if } i = 2. \end{cases}$$

[1]There is an analogue of this theorem for functions on a manifold of arbitrary dimension p, and for more general situations.

With the classical notation $\frac{1}{2}g''_{xx}(0) = r$, $\frac{1}{2}g''_{xy}(0) = s$, $\frac{1}{2}g''_{yy}(0) = t$, the point 0 is nondegenerate if $rt - s^2 \neq 0$. Taylor's formula with integral remainder is

$$g(x,y) - g(0) = R(x,y)x^2 + 2S(x,y)xy + T(x,y)y^2,$$

where

$$R(x,y) = \int_0^1 (1 - t)\, g''_{xx}(tx)dt,\ R(0) = r,$$

$$S(x,y) = \int_0^1 (1 - t)\, g''_{xy}(tx)dt,\ S(0) = s, \text{ and}$$

$$T(x,y) = \int_0^1 (1 - t)\, g''_{yy}(tx)dt,\ T(0) = t.$$

Case 1: $rt - s^2 > 0$, r and $t > 0$. In a neighborhood of 0, we have R and T > 0 and

$$g(x,y) - g(0) = R(x,y)\left(x + y\frac{S(x,y)}{R(x,y)}\right)^2 + y^2\frac{R(x,y)\ T(x,y) - S(x,y)^2}{R(x,y)}.$$

Setting

$$\begin{cases} X = \sqrt{R}\left(x + y\frac{S}{R}\right), \\ Y = y\sqrt{\frac{RT - S^2}{R}}, \end{cases}$$

we have $g(x,y) - g(0) = X^2 + Y^2$. Hence, the index of the critical point is 0. The pair (X,Y) is a system of local coordinates if the Jacobian matrix $\frac{D(X,Y)}{D(x,y)}$ has rank 2 at the origin. But at the origin we have

$$\frac{\partial X}{\partial x} = \sqrt{r},\ \frac{\partial Y}{\partial x} = 0,\ \frac{\partial Y}{\partial y} = \sqrt{\frac{rt - s^2}{r}}$$

and the Jacobian determinant is $\sqrt{rt - s^2} \neq 0$.

Case 2: $rt - s^2 > 0$, r and $t < 0$. This is reduced to Case 1 by considering -g. Hence, $g(x,y) - g(0) = -X^2 - Y^2$ and the index of the critical point is 2.

Case 3. $rt - s^2 < 0$. We reason as in Case 1 if r or t is not zero. Otherwise, we reduce to this hypothesis by a linear change of coordinates. We then find $g(x,y) - g(0) = X^2 - Y^2$ and the index of the critical point is 1.

Study of models. Let c be a nondegenerate critical point of f. By Theorem 2, there are local coordinates (X,Y) in a neighborhood of c. With respect to (X,Y), f is expressed as the sum of a constant and a quadratic form. The model functions are

$$g_0(X,Y) = X^2 + Y^2,$$
$$g_1(X,Y) = X^2 - Y^2, \text{ and}$$
$$g_2(X,Y) = - X^2 - Y^2.$$

The representative surfaces of the functions $Z = g_i(X,Y)$, $i = 0,1,2$, are respectively a paraboloid of revolution about the Z-axis, an equilateral hyperbolic paraboloid with principal direction $Z = 0$, and a paraboloid of revolution about the Z-axis.

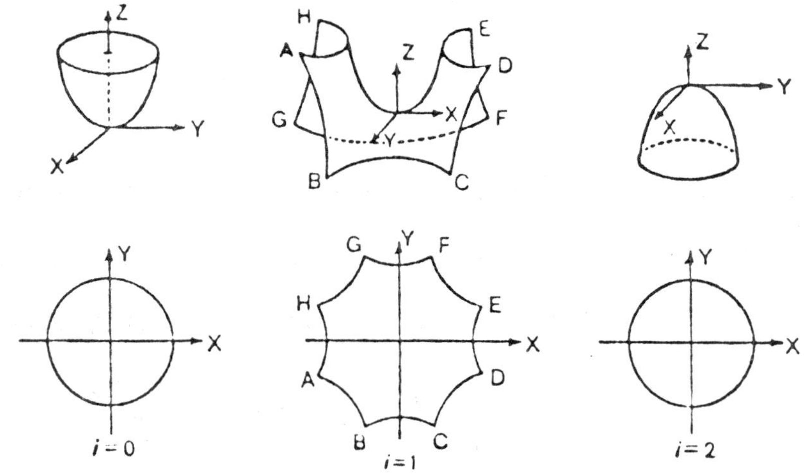

A *model neighborhood* for the function g_i is a neighborhood $U(s)$, $s > 0$, of 0 in R^2 defined as follows:

$i = 0$. $U(s)$ is a disk of radius $\sqrt{s}$. It is the set of points where $g_0(X,Y) \leqq s$, and is bounded by the projection of the curve $g_0(X,Y) = s$.

$i = 1$. The level lines of g_1 are the hyperbolas $X^2 - Y^2 =$ constant, their orthogonal trajectories are the hyperbolas $XY =$ constant. The neighborhood $U(s)$ is the set of points where $|X^2 - Y^2| \leqq s$ and $|XY| \leqq s$. It is bounded by the following arcs:

BC and FG of the hyperbola $X^2 - Y^2 = -s$ (lower level line),

AH and DE of the hyperbola $X^2 - Y^2 = s$ (upper level line),

AB and EF of the hyperbola $XY = s$,

CD and GH of the hyperbola $XY = -s$.

It is clearly homeomorphic to an octagon.

$i = 2$. $U(s)$ is the disk of radius $\sqrt{s}$. It is the set of points where $g_2(X,Y) \geqq -s$, and is bounded by the level line $g_2(X,Y) = -s$.

Canonical neighborhood. Let c be a nondegenerate critical point of f: $M \to R$. A *canonical neighborhood* U of c is the image of a parametrization ϕ: $U(s) \to M$ of M centered at c for which $f(\phi(X,Y)) = f(c) + g_i(X,Y)$ where i is the index of the critical point c. Theorem 2 insures the existence of the canonical neighborhoods.

COROLLARY. *The nondegenerate critical points are isolated critical points.*

For $i = 0, 1, 2$, we could verify that 0 is the only critical point of the function g_i. From this it follows that c is the only critical point of f in a canonical neighborhood U of c.

CHAPTER IV

MORSE FUNCTIONS ON SURFACES

In Chapter V we will see that specifying a Morse function (see definition below) on a compact surface M enables us to reconstruct the global topological structure of the space M--that is, to determine M to within a homeomorphism. The present chapter is devoted to the study of the behavior of a Morse function both in a neighborhood of a regular point and in a neighborhood of a critical point. This theory could be developed for manifolds of arbitrary dimension p, but this discussion will be limited to the case of surfaces. In this chapter, M will always denote a compact surface.

§1. MORSE FUNCTIONS ON A COMPACT SURFACE

Let f: $M \to R$ be a (real-valued) function of class C^r ($r \geqq 3$) all of whose critical points are nondegenerate. It follows from the corollary to Theorem 2 (Chapter III, §5) that such a function has a finite number of critical points. Indeed, since M is compact, if there were an infinite number of critical points, there would be a point $x \in M$ which would be an accumulation point of the critical points. At such a point $Df(x) = 0$, since Df is continuous and x would be a critical point. But that is impossible since from the corollary to Theorem 2 (Ch. III. §5) the critical points of f are isolated.

A *Morse function* f: $M \to R$ is a function all of whose critical points are nondegenerate and for which all the corresponding critical values (finite in number) are distinct. We assume the following theorem.[1]

THEOREM 1 (Morse). *Let* M *be a compact surface and let* g: $M \to R$ *be a continuous function. For each* $\varepsilon > 0$, *there exists a Morse function*[2] f: $M \to R$ *such that* $\| f - g \| < \varepsilon$.

This is the first global theorem that we have encountered. Chapter III dealt only with local properties of manifolds. For the case of surfaces, this theorem is a rather strong global property from which we can deduce the classification of all compact surfaces.

§2. VECTOR FIELDS AND ONE-PARAMETER GROUPS OF DIFFEOMORPHISMS

A *one-parameter group of diffeomorphisms* of class C^r on a manifold M is a mapping ϕ: $R \times M \to M$ (we write $\phi_t(x)$ instead of $\phi(t,x)$) such that:

i) For each t, the mapping $\phi_t: M \to M$ is a C^r-diffeomorphism.

ii) For all s and t, $\phi_{s+t} = \phi_t \circ \phi_s$, in particular, $\phi_0 = \mathrm{id}(M)$.

For $x \in M$, the velocity vector at the point $t = 0$ of the differentiable path $t \mapsto \phi_t(x)$ is a vector:

$$X(x) = [\frac{\partial}{\partial t} \phi_t(x)]_{t=0},$$

which is tangent to M at the point x. Its local coordinates are

---

[1]There is an "elementary" proof of this theorem in J. MILNOR, Morse theory, Notes by M. SPIVAK and R. WELLS, Princeton University Press, 1963.

[2]In the following, if the class of a differentiable function is not specified, it is assumed to be the class C^∞.

functions of class C^{r-1} of x. Such a specification, for all $x \in M$, of a tangent vector to M at x, as a function of class C^{r-1} of x, is a *tangent vector field* to M, of class C^{r-1}. For the case in which we are interested, we say that the field $X(x)$ generates the group with a parameter ϕ. Note that $X(\phi_t(x)) = [\frac{\partial}{\partial u} \phi_u(x)]_{t=u}$ since by i), if $u = t + h$, then

$$\frac{\partial}{\partial u} \phi_u(x) = \frac{\partial}{\partial h} \phi_{t+h}(x) = \frac{\partial}{\partial h} \phi_h(\phi_t(x)).$$

THEOREM 2. *Let* M *be a compact surface and let* X *be a tangent vector field to* M *of class* C^r. *There is a unique one-parameter group of diffeomorphisms of* M *generated by* X *which is of class* C^r.

First, we seek, for each $x \in M$, a parametrized curve on M of the form $y = \phi_t(x)$ such that:

$$\begin{cases} \frac{d}{dt}(\phi_t(x)) = X(\phi_t(x)) & (1) \\ \phi_0(x) = x. & (2) \end{cases}$$

With the local coordinates (x_1, x_2) on M, for which

$$X(x) = X_1(x_1,x_2)\frac{\partial}{\partial x_1} + X_2(x_1,x_2)\frac{\partial}{\partial x_2},$$

we seek two functions $y_1(t)$, $y_2(t)$ which are solutions of the system:

$$\frac{dx_1}{dt} = X_1(y_1,y_2), \quad \frac{dx_2}{dt} = X_2(y_1,y_2),$$

satisfying $y_1(0) = x_1$ and $y_2(0) = x_2$. For such a system the following local existence and uniqueness theorem for solutions is applicable.

THEOREM 3. *For each point* $x_0 \in M$, *there is a neighborhood* U *of* x_0 *in* M *and a number* $\varepsilon > 0$ *such that the equation (1,2) has a unique solution* $y = \phi_t(x)$ *defined for* $|t| < \varepsilon$ *and for all* $x \in U$, *and of class* C^r *in* (t,x).

Take a finite covering of the compact surface M by open sets U, obtained by Theorem 3, and let η be the smallest of the corresponding ε's. Then there is a mapping ϕ: $]-\eta,\eta[\times M \to M$ of class C^r defined by $(t,x) \mapsto \phi_t(x)$.

If $|t|$, and $|s|$ and $|s + t|$ are all less than η, we have $\phi_{t+s}(x) = \phi_t(\phi_s(x))$. In fact, assuming s to be constant, $\phi_{t+s}(x)$ and $\phi_t(\phi_s(x))$ are solutions of (1) which, for $t = 0$, equal $\phi_s(x)$, and hence by Theorem 3 they are the same. It follows from $\phi_0(x) = x$ that $\phi_t \circ \phi_{-t} = \phi_{-t} \circ \phi_t = \mathrm{id}(M)$ and that ϕ_t is a diffeomorphism. We have an "embryo of a one-parameter group." It remains to extend it to all the values of $t \in R$. To this end set

$$\phi_{nt} = \phi_t \circ \phi_t \circ \cdots \circ \phi_t \ (n \text{ times})$$

and verify that we have a one-parameter group.

§3. REGULAR VALUES OF A MORSE FUNCTION

Let f: $M \to R$ be a Morse function. If a is a regular (i.e., non-critical) value, denote by V(a) the curve (Chapter III, §4, Proposition 1) $f^{-1}(a)$, the *level curve* at a of the function f. We also denote by $M(a) = f^{-1}(]-\infty,a])$ the set of points of M such that $f(x) \leq a$. The interior of M(a) is the set of points where $f(x) < a$; its boundary is V(a). If $b > a$ is another regular value, denote by $W(a,b)$ the compact set $M(b)\backslash \mathrm{int}(M(a)) = f^{-1}([a,b])$.

PROPOSITION 1. *Suppose the Morse function* $f\colon M \to R$ *has no critical value between* a *and* b. *Then* $V(a)$ *is diffeomorphic to* $V(b)$, $M(a)$ *is diffeomorphic to* $M(b)$, *and* $W(a,b)$ *is diffeomorphic to* $V(a) \times [a,b]$.

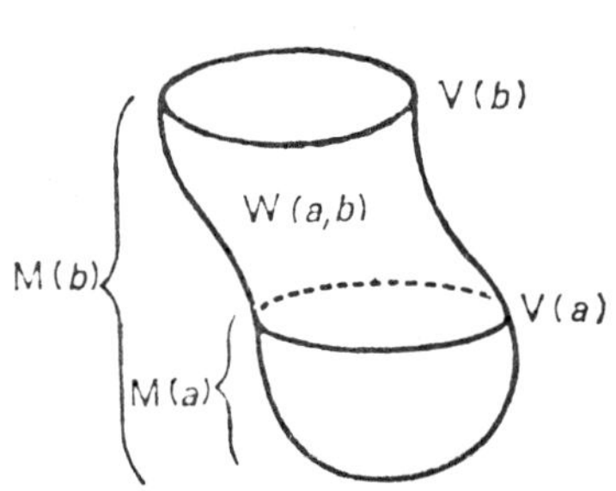

In order for the last two assertions to be meaningful, we should generalize the concept of differentiable mapping and manifold to *manifolds with boundary*. However, we shall be content with finding a homeomorphism from M(a) onto M(b) which induces a diffeomorphism of the interiors and a diffeomorphism of the boundaries (which are curves).

For each non-critical point $x \in M$, choose a tangent vector $X(x) \in T_x(M)$, transversal to the level curve $f(x) = constant$ and directed in the sense of increasing f, that is, a vector $X(x) = \alpha\frac{\partial}{\partial u} + \beta\frac{\partial}{\partial v}$ such that

$$Df.X(x) = \alpha\frac{\partial f}{\partial u} + \beta\frac{\partial f}{\partial v} > 0.$$

We can choose $X(x)$ to be differentiable in x, for instance, choose $X(x)$

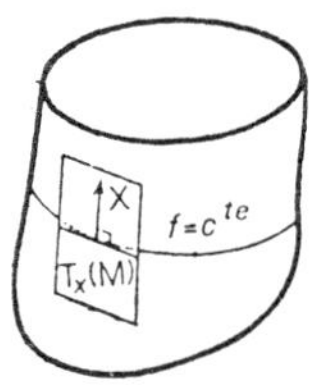

of length 1 and orthogonal to the level curves of f.

To complete the proof we shall use a differentiable function $\alpha\colon M \to R$ which is zero in a neighborhood of the critical points of f and equal to 1 outside a neighborhood of these critical points. One could be constructed as follows. In R^2, let ω be a differentiable function which equals 0 in the ball $D(\varepsilon)$ of radius ε and 1 outside the ball $D(2\varepsilon)$ of radius 2ε. If $\phi: D(2\varepsilon) \to M$ is a parametrization centered at the critical point c, then, for $x \in \phi(D(2\varepsilon))$, set

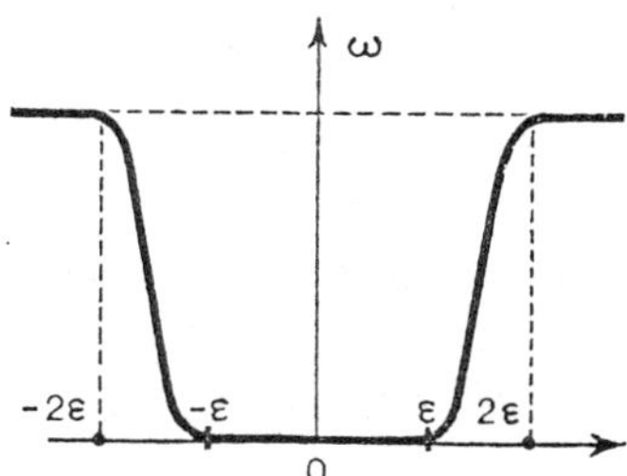

$\alpha(x) = \omega(\phi^{-1}(x))$. We may assume the neighborhoods $\phi(D(2\varepsilon))$ of the various (isolated) critical points to be disjoint, and extend α by 1 outside these neighborhoods.

Now define

$$Y(x) = \begin{cases} \dfrac{\alpha(x)}{Df.X(x)}.X(x) & \text{if } x \text{ is not a critical point,} \\ 0 & \text{if } x \text{ is a critical point.} \end{cases}$$

The vector field $Y(x)$ is differentiable and we have $Df.Y(x) = \alpha(x)$. Since there are no critical points in the compact set $W(a,b)$, it will be assumed that α equals 1 in this set. Let ϕ_t be a one-parameter group of diffeomorphisms of M associated with the vector field $Y(x)$. We have

$$\frac{d}{dt}(f(\phi_t(x))) = Df.Y(\phi_t(x)) = \alpha(\phi_t(x)).$$

The function $t \mapsto f(\phi_t(x))$ is linear with derivative 1 provided that $\alpha(\phi_t(x)) = 1$, which is the case for $f(\phi_t(x)) \in [a,b]$. In particular, the diffeomorphism ϕ_{b-a} sends M(a) onto M(b) and V(a) onto V(b). The mapping ψ: V(a) $\times$ $[a,b]$ $\to$ W(a,b), defined by $\psi(x,t) = \phi_{t-a}(x)$, is a "diffeomorphism" whose inverse is

$$\psi^{-1}(y) = (\phi_{a-f(y)}(y), f(y)).$$

Thus, when a varies continuously in the set of regular values of f, the manifolds M(a) and V(a) are invariant up to diffeomorphism. The curve V(a) is compact (closed in compact M). We assume that it is the union of a finite number of circles (up to diffeomorphism) which are its arcwise connected components (cf. Chapter V). The number of components of V(a) does not vary as long as a does not jump over a critical value.

§4. JUMPING OVER A CRITICAL VALUE

If a and b are regular values of f between which there is exactly one critical value $f(c)$ (which corresponds to a critical point c), then the manifolds M(a) and M(b) are not homeomorphic, and V(a) can be different from V(b). We are going to study the changes that take place when a critical value is jumped over. It follows from §3 that it may be assumed a and b are as close to $f(c)$ as we wish. Now take $a = f(c) - \varepsilon$, $b = f(c) + \varepsilon$ where we may pick $\varepsilon > 0$ as small as we wish.

Case 1--index 0. *If c is a critical point of index* 0, *then* M(b) *is diffeomorphic to the disjoint union of* M(a) *and a disk* D (of dimension 2), *and* V(b) *is diffeomorphic to the union of* V(a) *and the circle* S (the boundary of the disk D).

Let U be a canonical neighborhood of c, bounded by the appropriate component of $f^{-1}(f(c) + 2\varepsilon)$. The intersection D = M(b) $\cap$ U is diffeomorphic to the disk $X^2 + Y^2 \leqq \varepsilon$ and its boundary $S = V(b) \cap U$ is diffeomorphic to the circle $X^2 + Y^2 = \varepsilon$, while $V(a) \cap U$ and $M(a) \cap U$ are empty.

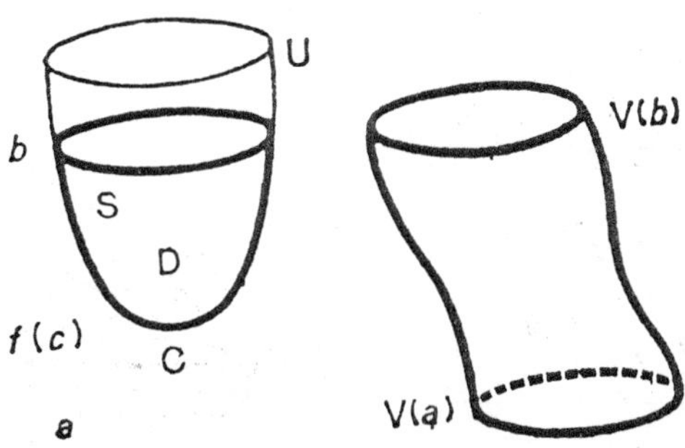

Then, by arguing as in §3, using a function α that is zero in a neighborhood of the disk D, we obtain a diffeomorphism ϕ_{a-b} from $M(b) \setminus D$ onto $M(a)$ which induces a diffeomorphism from $V(b) \setminus S$ onto $V(a)$.

Case 2--index 2. *If c is a critical point of index* 2, *then* M(*b*) *is homeomorphic to the space obtained by gluing a disk* D *to* M(*a*) *along a* (circular) *component* S *of* V(*a*), *and* V(*b*) *is diffeomorphic to* $V(a) \setminus S$.

It suffices to apply Case 1 to the function $-f$ by interchanging the roles of a and b.

Case 3--index 1. *If c is a critical point of index 1, then* M(*b*) *is homeomorphic to the space obtained by gluing to* M(*a*) *a rectangle with two of its opposite sides going to two disjoint segments* I *and* J *contained in* V(*a*). *The transformation of* V(*a*) *into* V(*b*) *depends on whether the number of* (circular) *connected components of V*(*b*) *differ from that of* V(*a*) *by -1, 0 or 1.*

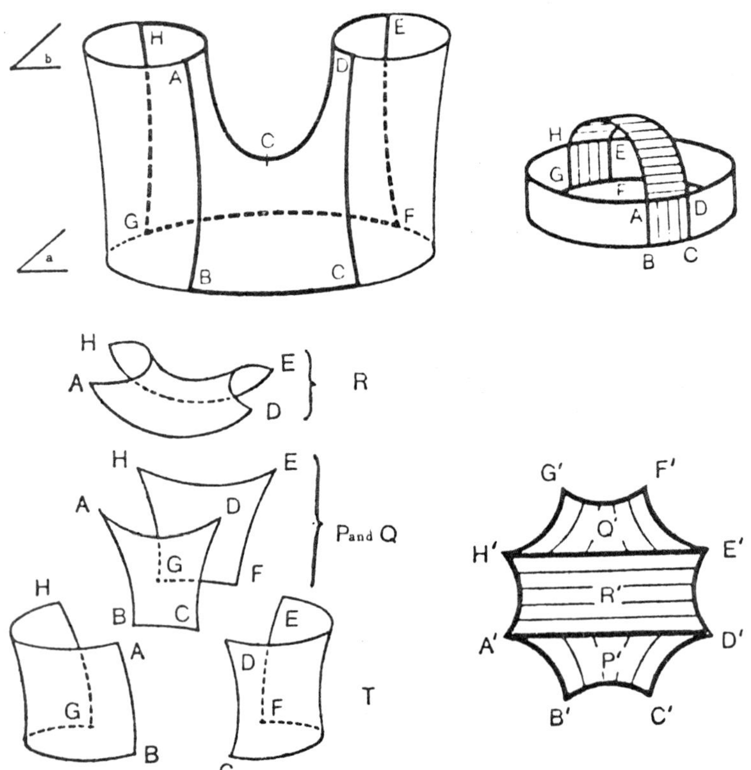

Let U' = $\phi(U(2\varepsilon))$ be a canonical neighborhood of c and let U = $\phi(U(\varepsilon))$ be a smaller canonical neighborhood, contained between the levels $a = f(c) - \varepsilon$ and $b = f(c) + \varepsilon$. The space M(b) is obtained by gluing W(a,b) to M(a) along V(a). The space W(a,b) is obtained by the gluing of U and T = $\overline{W(a,b)\backslash U}$. Finally, we shall decompose U into three pieces P, Q and R, and it is from this that M(b) can be reconstructed using many gluings. The intersection of U and V(a) consists of two segments I = BC and J = FG. The intersection of T and V(a) is the closure K of $V(a)\backslash(I \cup J)$.

We shall first show that T is homeomorphic to K × [a,b]. This is again done with the aid of a one-parameter group of diffeomorphisms ϕ_t of M. Then proceed as in §3 with a function α that is equal to zero in a neighborhood of c and is equal to 1 in T. We want the mapping $\psi: K \times [a,b] \to M$, defined by $\psi(x,t) = \phi_{t-a}(x)$, to be a homeomorphism on T.

This happens if the integral curves of the field $X(x)$ emanating from the points $x \in T$ remain in T between the levels a and b. That is equivalent to having the integral curves emanating from the points of U remain in U between the levels a and b.

Now construct a field $X(x)$ having the desired property. Let $Z(x)$ be the field on $U' = \phi(U(2\varepsilon))$ which is the image under $D\phi$ of the field $(X,-Y)$ of R^2. It is zero at c and elsewhere is transversal to the level lines and directed along increasing f (the verification is immediate in the chart ϕ^{-1}). Its integral curves are the images under ϕ of the hyperbolas XY = constant. Those level curves that emanate from a point of U are entirely in U between the levels a and b. Let $\beta: M \to R$ be a differentiable function equal to 1 in U and to 0 outside of $\phi(U(\frac{3}{2}\varepsilon))$ (construct in the chart and extend by 0 to M). Finally, let $X_0(x)$ be an arbitrary field on M, transversal to the level lines of f, and defined everywhere on M except at critical points. We set

$$X(x) = \beta(x)Z(x) + (1 - \beta(x))X_0(x) \text{ if } x \in U', \text{ and}$$

$$X(x) = X_0(x) \quad \text{if } x \notin U'.$$

It is clear that $X(x)$ is transversal to the level lines of f except at critical points. In U, we have $\beta = 1$, hence $X(x) = Z(x)$ and $X(x)$ has the required property.

Cut the model $U(\varepsilon)$ by two horizontals into three parts $P' = (A'B'C'D')$, $Q' = (E'F'G'H')$, and $R' = (A'D'E'H')$ (see the figure on p. 60), all three of which are homeomorphic to rectangles. Let P, Q, and R be their images in M under the parametrization ϕ. The space $M(b)$ is the result of the gluing of $M(a)$, T, P, Q and R; the figure on page 60 is more useful than a long explanation.

The spaces P and Q are homeomorphic to rectangles, for example, $I \times [a,b]$ and $J \times [a,b]$, respectively, so that these homeomorphisms glue (see Chapter II, §5) to the homeomorphism $\psi^{-1}: T \to K \times [a,b]$ to yield a homeomorphism from $T \cup P \cup Q$ onto $(K \cup I \cup U) \times [a,b] = V(a) \times [a,b]$. From this it follows that $M(b) \setminus \mathrm{int}(R) = M(a) \cup T \cup P \cup Q$ is obtained by gluing the cylinder $V(a) \times [a,b]$ to $M(a)$ along $V(a)$. Hence, this space is homeomorphic to $M(a)$ (see §3). Moreover, $M(b)$ is homeomorphic to the gluing together of R and $M(a)$, and is obtained by gluing AD onto I and EH onto J, which was to be proved.

Differences between $V(a)$ and $V(b)$. From the preceding constructions starting with $V(a)$, we obtain $V(b)$ to within homeomorphism, by means of the following procedure: remove from $V(a)$ the interiors of two small disjoint segments $I = BC$ and $J = FG$, and then glue two segments BG and CF to the space K obtained. The transformation described depends on the relative positions of B, C, F and G on $V(a)$ (see figure).

Case 1: *I and J are not in the same connected component of $V(a)$.* Let S be the component of I and S' that of J. The space $\overline{S \setminus I}$ is a segment BC, the space $\overline{S' \setminus J}$ is a segment FG; when we glue them together by the segments BG and CF we obtain a space homeomorphic to a circle S''. The space $V(b)$ is obtained starting with $V(a)$ and replacing $S \cup S'$ by S''.

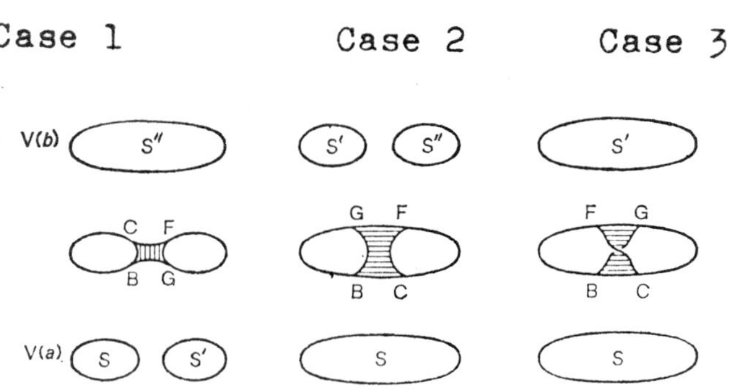

Case 2: *I and J are in the same connected component S of V(a) and the arcs* $\widehat{BC}$ = I *and* $\widehat{FG}$ = J *have the same orientation on* S. The space $\overline{S\backslash(I\cup J)}$ is the union of the two segments BG and CF. When we glue two new segments BG and CF, we obtain two circles S' and S". The space V(*b*) is obtained by starting with V(*a*) and replacing S by S'∪ S".

Case 3: *I and J are in the same connected component S of V(a) and the arcs* $\widehat{BC}$ = I *and* $\widehat{FG}$ = J *have opposite orientations on* S. The space $\overline{S\backslash(I\cup J)}$ is the union of two segments BF and CG. When we glue two segments BG and CF, we obtain a circle S'. The space V(*b*) is obtained by starting with V(a) and replacing S by S'.

All three situations will actually occur in the following.

§5. MODIFICATION OF A MORSE FUNCTION IN A CANONICAL NEIGHBORHOOD

We shall show that, if U is a canonical neighborhood of a critical point c of a Morse function f, then there exists another Morse function g that coincides with f outside of U and having only the critical point c in U, but with a critical value $g(c)$ which can be prescribed between certain limits. We obtain g by a simple modification of the function f in U.

PROPOSITION 2. *Let c be a critical point of index 0 of the Morse function f and let U be a canonical neighborhood of c bounded by the level curve $f(x) = a$. For each $b < a$, there is a Morse function g having the same critical points as f each with the same index, g coincides with f outside of U, and $g(c) = b$.*

There is a differentiable function with strictly positive derivative $\alpha\colon [f(c),a] \to R$ such that $\alpha(f(c)) = b$ and $\alpha(t) = t$ if t is close to a. If ϕ is the parametrization of the canonical neighborhood U, then for $x \in$ U we set $g(x) = \alpha(f(c) + g_0(\phi^{-1}(x)))$, where g_0 is the model function $X^2 + Y^2$. The only critical point of g in U is c, since the drivative of α is not zero. Now $g(c) = \alpha(f(c)) = b$, and when $\alpha(t) = t$, we have $g(x) = f(c) + g\,(\phi^{-1}(x)) = f(x)$. This is the case in a neighborhood of the boundary of U, which proves that the function g on U glues to f on M\U to yield the Morse function g sought.

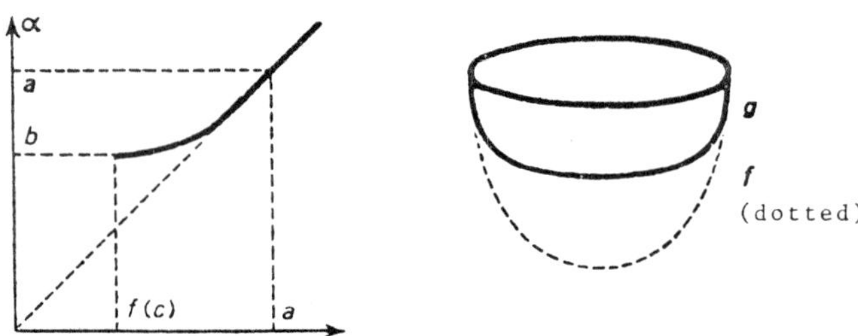

For a point of index 2, Proposition 2 applied to the function $-f$ shows that $g(c)$ can be chosen arbitrarily large, and strictly higher than the level of the boundary of a canonical neighborhood.

PROPOSITION 3. *Let c be a critical point of index 1 of the Morse function f and let* U *be a canonical neighborhood of c bounded at the lower level by $f(x) = a$ and at the upper level by $f(x) = b$. For each $d \in]a,b[$, there is a Morse function g, having the same critical points as f, each with the same index, g coincides with f outside of* U, *and $g(c) = d$.*

As before, it suffices to alter the model function $g_1 = X^2 - Y^2$ in the interior of the model neighborhood. Assume that the model neighborhood is U(1). We will show that there is a function $h(X,Y)$, coinciding with $g_1(X,Y)$ in a neighborhood of the boundary of U(1), whose only critical point is 0 and whose corresponding critical value is $h(0) = -\beta \in]-1,0]$ (if β is negative, just reverse the roles of X and Y).

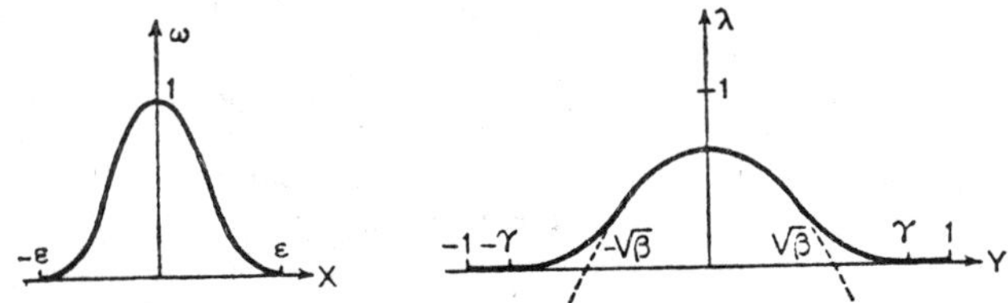

Let ω: $R \to R$ be a positive differentiable function with support in $[-\varepsilon,\varepsilon]$ with $\omega(0) = 1$ and $X.\omega'(X) \leqq 0$. Let λ: $R \to R$ be a positive differentiable function with support in $[-\gamma,\gamma]$, where $\gamma \in]\sqrt{\beta},1[$ is such that $\lambda(0) = \beta$ and whose derivative $\lambda'(Y)$ satisfies $2Y + \lambda'(Y) > 0$ if $Y > 0$ and $2Y + \lambda'(Y) < 0$ if $Y < 0$. For $\eta > 0$, the function $\mu(Y) = \beta - Y^2 + \eta Y^2$ satisfies $\mu(0) = \beta$ and the above conditions on the derivative. If η is sufficiently small, $\mu(Y)$ is negative for $Y = \pm\gamma$. To obtain λ, it suffices to restrict the function μ to a neighborhood of its zeros $Y = \pm\sqrt{\frac{\beta}{1-\eta}}$.

If ε is sufficiently small, the rectangle $[-\varepsilon,\varepsilon] \times [-\gamma,\gamma]$ is contained in the interior of $U(1)$ so that the function

$$h(X,Y) = X^2 - Y^2 - \omega(X)\lambda(Y)$$

coincides with g_1 in a neighborhood of the boundary of $U(1)$. We have $h(0) = -\omega(0)\lambda(0) = -\beta$. It remains to show that 0 is the only critical point of h in $U(1)$. The partial derivatives are

$$h'_X = 2X - \omega'(X)\lambda(Y), \text{ and}$$

$$h'_Y = -2Y - \omega(X)\lambda'(Y).$$

From the fact that $\lambda(Y) \geqq 0$ and that $\omega'(X)$ has the same sign as $-X$, it follows that the derivative h'_X vanishes only if $X = 0$. In this case, we have $h'_Y = -2Y - \lambda'(Y)$. By construction of the function λ, the derivative h'_Y vanishes only for $Y = 0$. This completes the proof of Proposition 3.

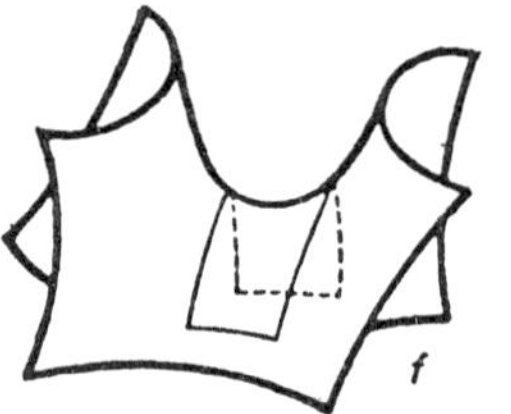

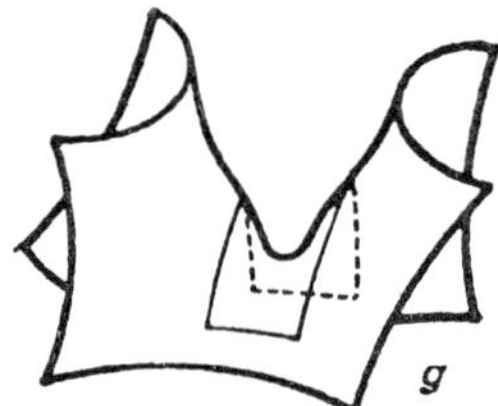

A Morse function is said to be *ordered* if the indices of the critical points are an increasing function of the critical values (the critical values of index 0 are less than the critical values of index 1, etc.). As a consequence of Proposition 2 and Theorem 1, we have the following theorem.

THEOREM 4. *On a compact surface there exists an ordered Morse function.*

Let f be a Morse function on M (Theorem 1) and let a and b be lower and upper bounds of the critical values (finite in number) of index 1. By Proposiiton 2, we can find a function g having the same critical points as f, each with the same index, whose critical values of index 0 are less than a and whose critical values of index 2 are greater than b.

CHAPTER V

THE CLASSIFICATION OF SURFACES

Let M be a compact connected differentiable manifold of dimension 2 (i.e., a surface). By studying a Morse function on M, we shall show that M is homeomorphic to one of the spaces T_n $(n \geqq 0)$ or U_n $(n \geqq 1)$ described in Chapter II. Conversely, each of these spaces is homeomorphic to a differentiable manifold so that we have a complete classification (to within homeomorphism) of all the compact connected differentiable surfaces. There is no analogue[1] for manifolds of dimension $p \geqq 3$.

§1. THE CLASSIFICATION OF CURVES

We shall first show that the study of compact manifolds reduces to the study of compact connected manifolds.

THEOREM 1. *A compact topological manifold of dimension p has a finite number of arcwise connected components. These components are compact manifolds of dimension p.*

The arcwise connected components of a manifold are open (see Chapter I, Proposition 4). If there were an infinite number of them, then by choosing a point in each component we could construct a sequence

---

[1]Cf. A. MARKOV, *Insolubility of the problem of homeomorphy*, Proc. Intern. Congress of Mathematicians, 1958, Cambridge University Press, pp. 300-306.

not having an accumulation point (every point of the manifold has a neighborhood, its arcwise connected component, which contains only one point of the sequence). Hence there are a finite number of components. Each component being open, is a manifold of dimension p and being closed in a compact space, is compact.

THEOREM 2. *Every compact connected differentiable manifold of dimension* 1 *is homeomorphic to a circle.*[1]

Let M be a compact connected differentiable curve and let $f\colon M \to R$ be an ordered Morse function (see Chapter IV, Theorem 4). Since M is compact, the function f has a minimum c which is a critical point of index 0, and a maximum c' which is a critical point of index 1.

We first assume that these are the only critical points of f, and let a be a level such that $f(c) < a < f(c')$. The space $M(a) = f^{-1}(]-\infty, a])$ is homeomorphic to a closed segment (see Chapter IV, Proposition 1 and §4) and its boundary $V(a) = f^{-1}(a)$ consists of two points A and B. Similarly, the space $M'(a) = f^{-1}([a, +\infty[)$ is homeomorphic to a closed segment and has the same boundary. The space M is obtained by gluing M'(a) and M(a) along V(a). We remark that the circle is also obtained by gluing together two segments. By the gluing together of homeomorphisms (see Chapter II, §5), it follows from this that M is homeomorphic to a circle.

In the general case, where there are n critical points of index 0, if a is a level separating the levels of critical points of index 0 from those of index 1, then the space M(a) is homeomorphic to the disjoint union of n segments (see Chapter IV, §4) and V(a) consists of $2n$ points.

---

[1]It is even true that every compact connected *topological* manifold of dimension 1 is homeomorphic to a circle and that every compact connected differentiable curve is diffeomorphic to a circle.

From this it follows that M'(a) consists of n segments and that there are n critical points of index 1.

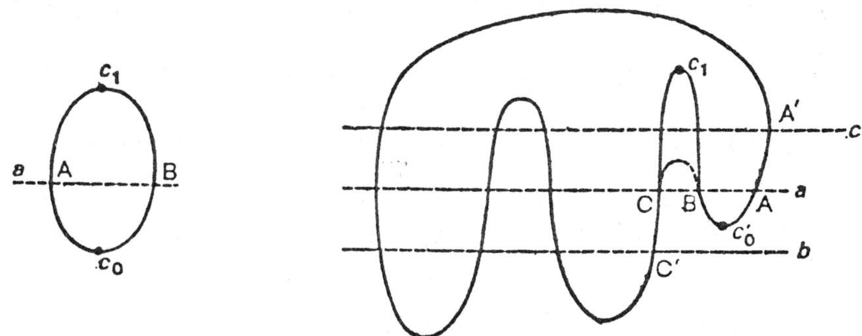

Let c_n be the critical point of index 0 having the highest level. It is situated on a component having endpoints A and B of M(a). In M'(a), the component D_1 which contains B contains a critical point of index 1, denoted by c'_n. The other endpoint C of D_1 is a point in level a distinct from A, for otherwise M would not be connected. By modifying f only on the segment D_1, we replace f by a Morse function f' for which c'_n is the critical point of index 1 that has the lowest level and $f'(c'_n) > a$ (see Chapter IV, §5, Proposition 2). Let b and c be levels such that $b < f'(c_n) < a < f'(c'_n) < c$ and such that c_n and c'_n are the only critical points of f' whose values are between b and c. We shall show that M(b) is homeomorphic to M(c). The space M(b) is homeomorphic to the union of (n - 1) segments. To obtain M(c), we slightly extend (n - 2) of the segments; as for the last one, we slightly extend one of its endpoints and in succession we glue to the other the segments C'C, CB, BA and AA' (see the figure). Consequently, M(c) is also homeomorphic to the union of (n - 2) segments.

The manifold M is obtained by gluing together M'(c) and M(c) (which is homeomorphic to M(b)). The critical points c_n and c_n' play no role[1]. The manifold M, obtained by gluing two families of (n - 1) segments is thus homeomorphic to a curve M' on which there exists a Morse function with (2n - 2) critical points (the gluing together of homeomorphisms). This shows, by induction on n, that M is homeomorphic to a circle.[2]

§2. PRELIMINARIES TO THE CLASSIFICATION OF SURFACES

We establish the classification of surfaces by a proof analogous to that of Theorem 2, but obviously more complicated since there are critical points of index 0, 1 and 2. Henceforth, M will denote a compact connected surface, and f: M → R an ordered Morse function. First we obtain an analogue of the elimination procedure used in the proof of Theorem 2.

Assume that between the levels b and c the function f has exactly two critical points c_0 and c_0' of index 0 and 1 respectively. We say that the critical points c_0 and c_0' are *in good position* if c_0 has a canonical neighborhood U_0 extending above level d and c_0' has a canonical neighborhood U_1 extending below level d so that in the level curve V(d) one of the two segments composing $U_1 \cap V(d)$ (and only one) is contained in the circle which constitutes $U_0 \cap V(d)$.

---

[1]We could prove that one can modify f' on the segment A'C', transforming it into a Morse function which no longer has a critical point on the segment A'C'.

[2]This easy proof has been given in detail in order to clarify the ideas in the two-dimensional analogue.

PROPOSITION 1. *With the above terminology, if the critical points* c_0 *and* c'_0 *are in good position, then the space* M(c) *is homeomorphic to the space* M(b).

By Chapter IV, §4, the space M(d) is diffeomorphic to the disjoint union of M(b) and a disk D. Similarly, the space M(c) is homeomorphic to the space obtained by gluing together M(b) and D with the aid of a rectangle R, by identifying one of the sides of the rectangle with a segment I contained in V(b) and the opposite side with a segment J contained in the boundary of D. It is easy to see that M(c) is homeomorphic to M(b).[1]

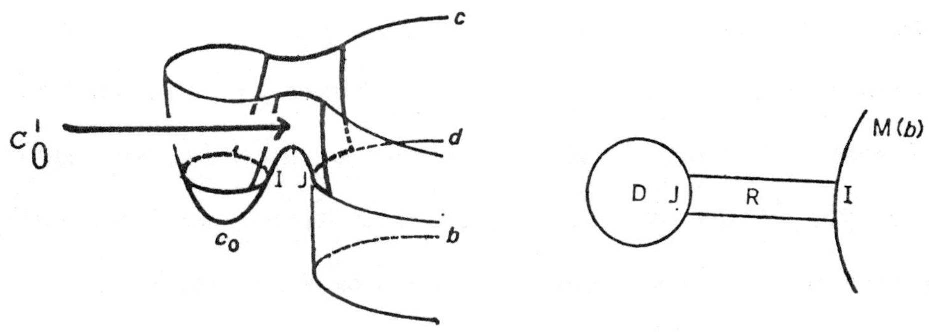

We now modify slightly the notion of a canonical neighborhood of a critical point of index 1. In §5 of Chapter III we considered the model neighborhoods U(s) defined by the inequalities $|X^2 - Y^2| \leqq s$ and $|XY| \leqq s$. We shall now assume as model neighborhoods the neighborhoods of 0 defined by $-r \leqq X^2 - Y^2 \leqq s$ and $|XY| \leqq t$, where $r, s, t > 0$. It is

[1]We could show that there exists a Morse function f' which coincides with f except for the level between b and c and which has no critical points in the set $f^{-1}([b,c])$. In other words, two critical points in good position can be eliminated without changing the other critical points.

easy to see that Proposition 3 of Chapter IV still holds with this new definition.

Let a be a level greater than that of all the critical points of index 0 and less than that of all the critical points of index 1. Similarly, let b be a level between the levels of the critical points of indices 1 and 2. The space $M(a) = f^{-1}(]-\infty,a])$ is a disjoint union of disks (see Chapter IV, §4) and each of them is a canonical neighborhood of a critical point of index 0. Similarly for $M'(b) = f^{-1}([b,+\infty[)$ and the critical points of index 2. For critical points of index 1, we shall assume the following result.

PROPOSITION 2. *For each critical point of index* 1, *there is a canonical neighborhood extending below level* a, *and above* b. *Furthermore, these canonical neighborhoods are pairwise disjoint.*[1]

Therefore, using Proposition 3 of Chapter IV, we can change the order of the values of f at critical points of index 1 by modifying f in those canonical neighborhoods obtained from Proposition 2.

§3. BEGINNING OF THE PROOF

We consider the compact connected surface M and the ordered Morse function f: M → R. The function f has $n(0)$, $n(1)$ and $n(2)$ critical

---

[1]The proof is not difficult, but tedious; here is the essence of it. Take "small" pairwise disjoint canonical neighborhoods of all the critical points. We try to extend them with the aid of a vector field analogous to the field X (of Chapter IV, §4, Case 3) for all canonical neighborhoods simultaneously. The extension terminates if the integral curves of X emanating from a canonical neighborhood U begin to penetrate into another canonical neighborhood U' between the levels a and b. We then show that one can always modify X so as to avoid this phenomenon.

points of index 0, 1 and 2, respectively. There is at least one critical point of index 0 (the minimum of f) and at least one critical point of index 2 (the maximum of f). As before denote by a and b the levels separating the critical points of index 0, 1 and 2. The space M(a) is the disjoint union of a family of disks D_i ($1 \leqq i \leqq n(0)$), canonical neighborhoods of the critical points c_i of index 0. The space M'(b) is the disjoint union of a family of disks D''_k ($1 \leqq k \leqq n(2)$), canonical neighborhoods of the points c''_k of index 2. We choose (see Proposition 2) *disjoint* canonical neighborhoods U_j ($1 \leqq j \leqq n(1)$) of the critical points c'_j of index 1 extending above level b and below level a. It follows from §4 of Chapter IV that the space M(b) is homeomorphic to the union $\hat{M}(b)$ of the D_i and the U_j; it suffices to apply the rule for jumping over a critical point of index 1 successively to each of the c'_j. In other words, M(b) is homeomorphic to the gluing together of $n(0)$ disks and $n(1)$ rectangles (the octagons U_j are homeomorphic to rectangles). We then obtain M by again gluing $n(2)$ disks. We are going to simplify this construction to the extent that one can recognize the result.

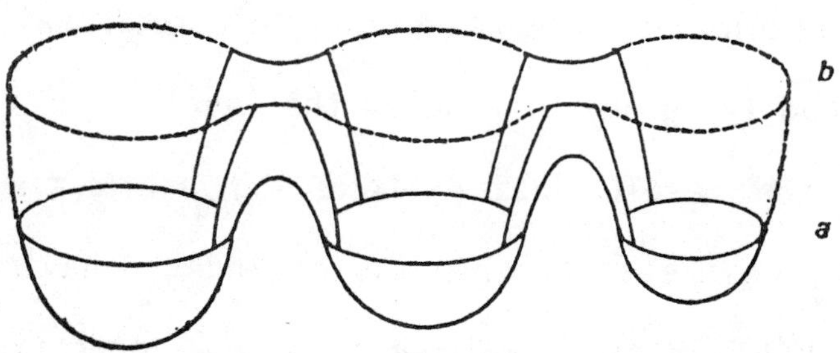

We first reduce it to a situation analogous to that of a function which would have only a single critical point of index 0 and a single critical point of index 2, that is, for which M could be constructed using rectangles and two disks.

LEMMA 1. *There exists on* M *a Morse function* g *having the same critical points as* f, *each with the same index, and two levels* c *and* d, *with* $c < d$, *such that:*

(i) The space $M(c) = g^{-1}(]-\infty,c])$ *contains* $n(0)$ *critical points of index* 0 *and* $n(0) - 1$ *critical points of index* 1; *moreover,* $M(c)$ *is homeomorphic to a disk.*

(ii) The space $M(d) = g^{-1}([d,+\infty[)$ *contains* $n(2)$ *critical points of index* 2 *and* $n(2) - 1$ *critical points of index* 1; *moreover,* $M(d)$ *is homeomorphic to a disk.*

(iii) The space $W(c,d) = g^{-1}([c,d])$ *contains* $n(1) - n(0) - n(2) + 2$ *critical points of index* 1.

In particular, this shows that $n(0) - n(1) + n(2) \leqq 2$.

Proof of the lemma. To prove the lemma, it suffices to show that, by modifying f only below the level b, we can find a function g and a level c satisfying the first assertion of the lemma. In fact, it then suffices to apply the procedure anew, interchanging the roles of the indices 0 and 2 (think of the function $-g$) to obtain the second assertion without modifying anything below the level c.

If $n(0) = 1$, the result holds with $f = g$ and $c = a$. We assume $n(0) \geqq 2$ and let c_1 be the critical point of index 0 having the highest level. The space $M(b)$ is the complement (in the connected manifold M) of $n(2)$ disjoint open disks. From this it easily follows that $M(b)$ is arcwise connected. Since $M(a)$ has several connected components D_i, there necessarily exists an octagon, U_1 for example, which joins D_1 to some other component, D_2 for example. The critical points c_1 and c_1' are in good position, because U_1 intersects two components of $V(a)$, provided there is no critical value in the interval $]f(c_1),f(c_1')[$. We can reduce the situation to this case by modifying f in the interior of U_1 (see Chapter IV, Proposition 3) and lowering $f(c_1')$ until it is smaller than the other critical values of index 1.

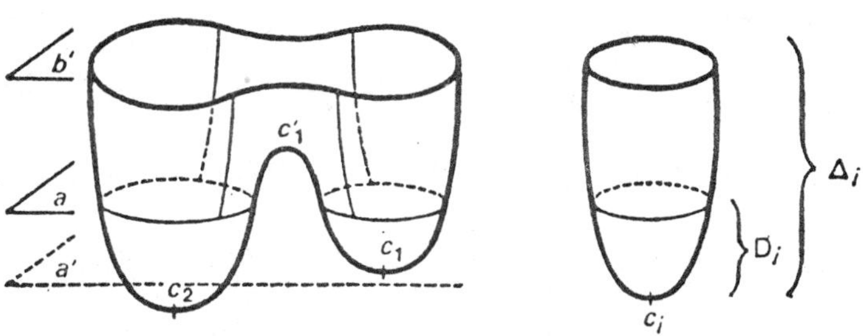

So, let a' and b' be two levels such that $f(c_1)$ and $f(c_1')$ are the only critical values in the interval $]a',b'[$. By Proposition 1, the space $M(b')$ is homeomorphic to the space $M(a')$ which consists of canonical neighborhoods of the points $c_2,c_3,\ldots,c_{n(0)}$. Therefore, the space $M(b')$ has $n(0) - 1$ components each homeomorphic to a disk. One of them, say Δ_1, contains the three critical points c_1,c_1' and c_2. The others, $\Delta_2,\ldots,\Delta_{n(0)}$, are the canonical neighborhoods of the points $c_2,\ldots,c_{n(0)}$.

In order to continue this procedure of reducing the number of disks, it then suffices to modify f in $\Delta_2,\ldots,\Delta_{n(0)}$ so that the critical values $f(c_2),\ldots,f(c_{n(0)})$ are greater than $f(c_1')$ (see Chapter IV, Proposition 2). If $n(0) - 1 \geqq 2$, the space $M(b)$ is connected, so there is an octagon U_j intersecting simultaneously Δ_3 (where c_3 is the highest point of index 0) and another disk Δ_i. We now proceed with c_3 and c_j' as before with c_1 and c_1'. By induction, this yields the lemma.

§4. ORIENTABLE AND NON-ORIENTABLE SURFACES

We say that one has oriented a differentiable surface if one has chosen in each tangent plane $T_a(M)$ an orientation which depends continuously on the point a. That is, if X and Y are two (continuous) fields of vectors tangent to M, each defined in a neighborhood of a, and if the vectors $X(a)$ and $Y(a)$ form a positively oriented frame in $T_a(M)$ the same is true of $X(x)$ and $Y(x)$ in $T_x(M)$ for x close to a. We say that a surface is *orientable* if it can be oriented.

Suppose that the surface M possesses a family of parametrizations $\phi_i: U_i \to M$ (where U_i is an open set in R^3 and i is an index ranging over a set I) which forms a complete system (every point of M is in the image of one of the parametrizations). Further, suppose that all the changes of charts $\theta_{ij} = \phi_i^{-1} \circ \phi_j$ are diffeomorphisms from an open set in R^2 which preserve orientation (positive Jacobian determinant). At every point a of M, the orientation of $T_a(M)$ defined by the frame $(\frac{\partial}{\partial x}, \frac{\partial}{\partial y})$ associated with a parametrization ϕ_i (cf. Chapter III, §3) is independent of ϕ_i $(i \in I)$ and depends continuously on a. We say that the surface M is *oriented by the system* (ϕ_i).

Conversely, if M is oriented, then for every point $a \in M$, there is a parametrization ϕ of a neighborhood of a such that $D\phi$ sends the canonical orientation of R^2 *into the orientation of* $T_x(M)$ (x close to a). The family of all these parametrizations as a ranges over M forms a complete system where the changes of parametrization obviously preserve the orientation.

Example 1. The interior of a Möbius strip is a non-orientable (non-compact) surface.

PROPOSITION 3. *A surface that contains a subspace diffeomorphic to the interior of a Möbius strip is non-orientable.*

This is obvious by Example 1.

PROPOSITION 4. *A surface in* R^3 *that admits a continuous field of nonzero normal vectors is orientable.*

Let $M \subset R^3$ be a surface. A *continuous field of normal vectors* is the specification for each $a \in M$ of a vector $N(a)$ with origin a that is normal to the tangent plane $T_a(M)$, and for which the specification depends continuously on a. It is clear that such a specification, with $N(a) \neq 0$, determines an orientation of $T_a(M)$ which depends continuously on a.

Remark 1. We shall show in Chapter VII that every *compact* surface in R^3 does have a continuous field of normal vectors and, consequently, is orientable.

We have seen (see Chapter IV, §4) that there are three possible changes in the level curves caused by jumping over a critical point of index 1. If we are in the first or the second case, where the number of components of the level curve changes by one, we say that the *critical point is of type I*, otherwise that it is *of type II*.

PROPOSITION 5. *Let* M *be a surface and let* f *be a Morse function on* M. *If* f *has a type II critical point of index* 1, *then* M *is non-orientable.*

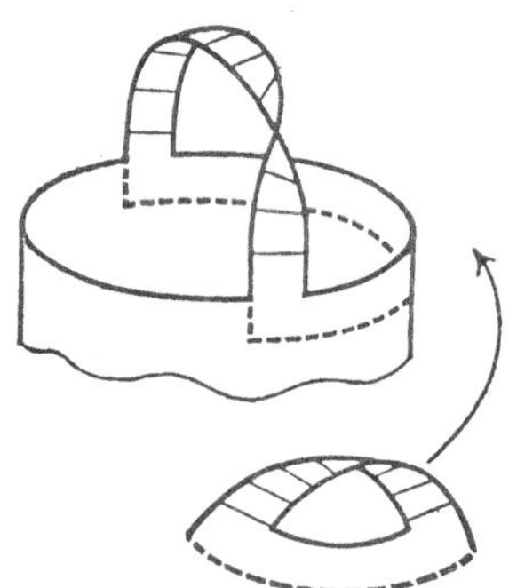

It suffices to prove (see Proposition 3) that M contains a part diffeomorphic to the interior of a Möbius strip. It is easy to see this in the figure. It is more difficult to explain, but does not pose any serious problem.

PROPOSITION 6. *Let* M *be a compact surface and let f be a Morse function on* M. *Let g be the associated Morse function from Lemma 1. Between the levels c and d, there are an even number of critical points of type I. All the critical points whose level is outside* $[c,d]$ *are of type I.*

The second assertion follows from the construction of the function g. The first assertion is a consequence of the fact that $V(c)$ and $V(d)$ each has a single component and on the other hand that the points of type I are those which change the parity of the number of components of the level curve.

§5. A SPECIAL CASE

Because of Lemma 1, we shall indulge in the following abuse of language. We use the notation f, a and b to designate g, c and d of the lemma and say that f has one critical point of index 0, that it has

$n(1) - n(0) - n(2) + 2$ critical points of index 1 and one critical point of index 2. From the point of view of the construction of M by gluings, this abuse of language is justified. Moreover, we could prove the existence of such a Morse function on M.

THEOREM 3. *If the function f has one critical point of index* 0, *one critical point of index* 2 *and q type II critical points of index* 1 *on the compact connected surface* M, *then* M *is homeomorphic to the space* U_q (see Chapter II, §7).

Recall that the space U_q is the iterated connected sum of q copies of the projective plane $P_2 = U_1$. For convenience, we set $U_0 = S^2$.

The case where $q = 0$. Note that, on the sphere $x^2 + y^2 + z^2 - 1 = 0$, the projection to the z-axis is a Morse function without a critical point of index 1. The sphere $S^2 = U_0$ is obtained by gluing together the two hemispheres $(z \geqq 0)$ and $(z \leqq 0)$ which are the canonical neighborhoods of the two critical points.

For the surface M, we consider a level c between the levels a and b. The surface M is obtained by the gluing together of M(c) (which is homeomorphic to M(a)) and M'(c) (which is homeomorphic to M'(b)). By the gluing together of homeomorphisms (see Chapter V, §7 below) it follows that M is homeomorphic to U_0.

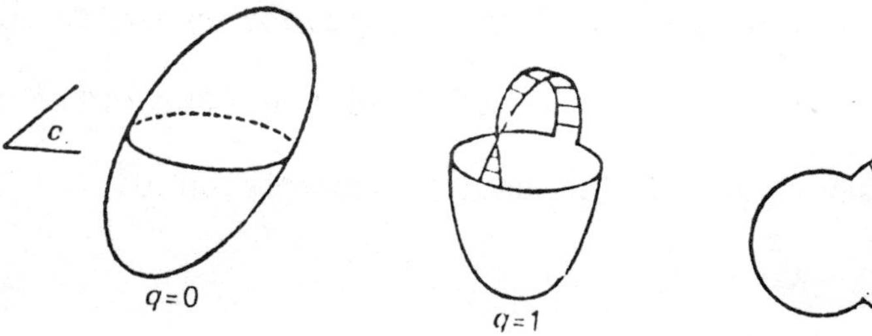

The case where $q = 1$. In this case, the space $M(b)$ is obtained by gluing a rectangle to a disk with non-compatible orientations. It is easy to see that $M(b)$ is homeomorphic to a Möbius strip. Then M, which is obtained by gluing $M(b)$ and the disk $M'(b)$ along their circular boundaries, is homeomorphic to the projective plane $P_2 = U_1$.

The case where $q \geqq 2$. We assume (for induction) that the theorem has been proved for a Morse function having $q - 1$ critical points of index 1. We shall now consider the case of q critical points, denoted by $c_1', \ldots, c_q'$, in increasing order of the critical values. Let c be a value between $f(c_1')$ and $f(c_2')$. It follows from the study of the case $q = 1$ that $M(c)$ is homeomorphic to a Möbius strip and it follows from the induction hypothesis that $M'(c)$ is homeomorphic to U_{q-1} with a disk removed (the function $-f$ has $q - 1$ critical points of index 1 below the level $-c$). By gluing these together, we see that M is homeomorphic to $U_1 \# U_{q-1} = U_q$.

§6. THE GENERAL CASE

THEOREM 4. *Every connected surface is homeomorphic to either a* U_q *(*$q \geqq 0$*) or a* T_p *(*$p \geqq 0$*).*

We are going to prove the following assertion by induction on q. *If the Morse function on* M *has one critical point of index* 0*, one critical point of index* 2 *and* q *critical points of index* 1*, then the space* M *is homeomorphic to* U_q *if* q *is odd, and homeomorphic to either* U_q *or to* T_p *if* $q = 2p$. We have already considered the case where $q = 0$ in §5, and also the case where $q = 1$ (since by Proposition 6, the critical point is of type II).

The case where $q = 2$. If the two critical points are of type II, M is homeomorphic to U_2 by Theorem 3. Otherwise, the two critical points are of type I (by Proposition 6). For the torus T_1 (see Chapter III, §4, Example 4), we have a Morse function that has two critical points of index 1, one critical point of index 0 and one critical point of index 2. Let a', c', b' be values that separate the four critical values. The space $T_1(c')$ is homeomorphic to a cylinder $S \times I$; we can also obtain it by gluing a rectangle to the disk $T_1(a')$ using two opposite sides with compatible orientations.

For the surface M, let c be a value between $f(c_1')$ and $f(c_2')$. The space $M(c)$ is homeomorphic to $T_1(c')$ inasmuch as one obtains it by gluing a rectangle to $M(c)$ with compatible orientations. Similarly, $M'(c)$ (homeomorphic to $T_1'(c')$) is homeomorphic to a cylinder $S \times I$. To obtain M, we glue together these cylinders. There are two ways of doing this; either with compatible orientations, in which case we obtain the torus T_1, or with opposite orientations, in which case we obtain the Klein bottle U_2.

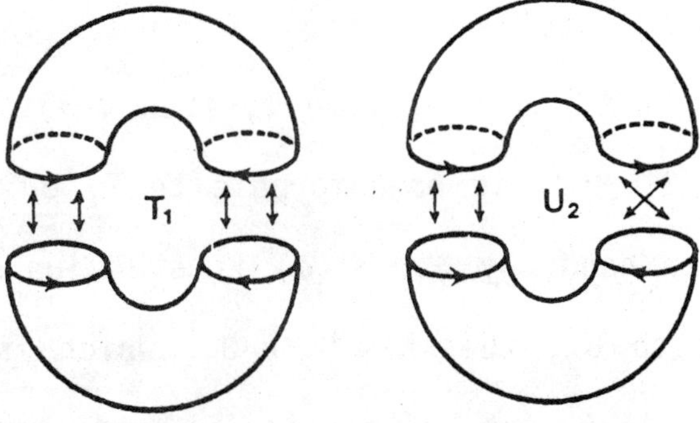

We obtain the same result by studying the space $M(b)$. This is obtained by gluing two rectangles to a disk. To glue the first, there is only one way, up to homeomorphism if the boundary of $M(c)$ is to be two circles. For the second, whose two opposite sides are each glued to one of these circles, there are two ways.

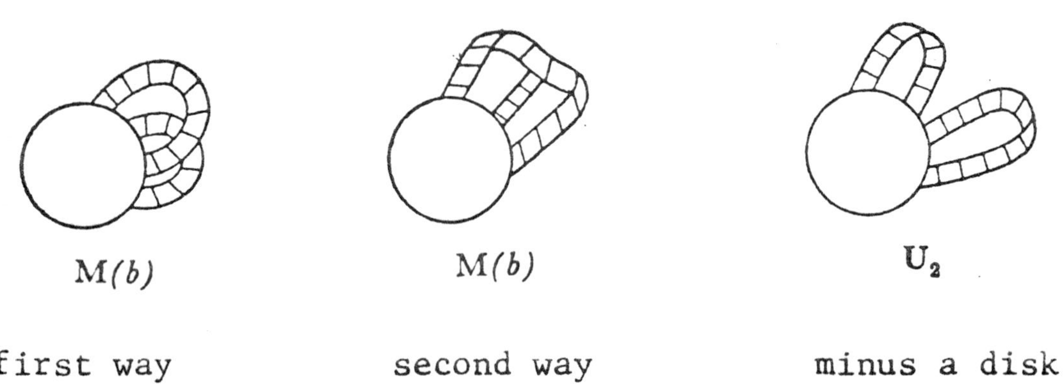

first way second way minus a disk

The first yields $M(b)$ which is homeomorphic to V_1. The second is homeomorphic to U_2 with a disk removed (see §5).

The case where $q = 3$. There can be three points of type II in which case M is homeomorphic to U_3 (Theorem 3), or one point of type II and two points of type I (see Proposition 6). In this case one of the points of type I is either the highest or the lowest. We may assume that it is the lowest point c_1' (changing f into $-f$ if necessary). Let c and d be levels such that $f(c_1') < c < f(c_2') < d < f(c_3')$.

First we assume that c_2' is of type I; then $V(d)$ has one component and by the case $q = 2$, $M(d)$ is homeomorphic to T_1 or U_2 with one disk removed. On the other hand, by §5, $M'(d)$ is a Möbius strip. Hence, if $M(d)$ is homeomorphic to U_2, then $M = U_2 \# U_1$, which is homeomorphic to U_3. We shall show that the situation is the same when $M(d)$ is homeomorphic to T_1. That is, we shall show that $T_1 \# U_1$ is homeomorphic to U_3. For this, in both cases we compare the spaces $M(b)$ which are obtained by gluing three rectangles to a disk. It suffices to show that they are homeomorphic, since then the spaces M, obtained by gluing a disk to $M(b)$ will also be homeomorphic. We see that the two spaces $M(b)$ are homeomorphic by studying the drawings below:

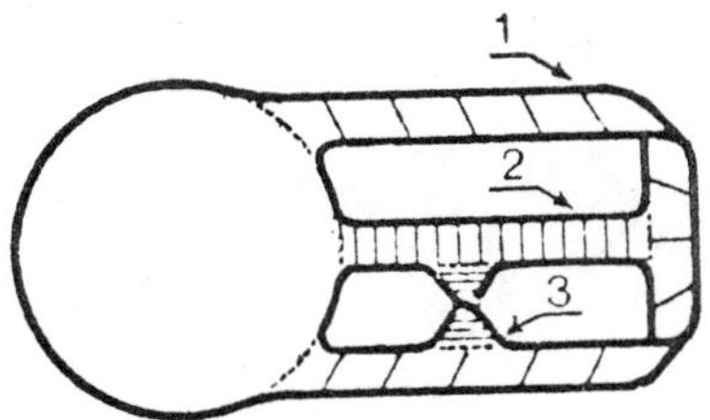

Case 1: $U_2 \# U_1$

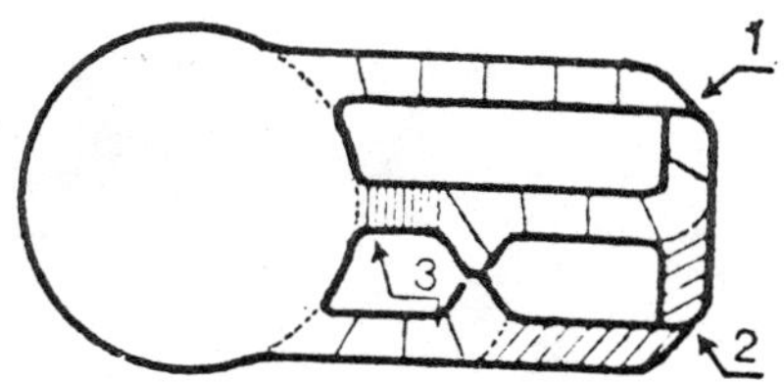

Case 2: $T_1 \# U_1$

We now show that if the point c_2' is of type II, then f can be modified in order to change this into the preceding case. The point c_3' is of type I and its canonical neighborhood U_3 intersects the two components of $V(d)$ since $V(b)$ has only one component. Hence, it also intersects the two components of $V(c)$. If we modify the function f on V_3 (see Chapter IV, Proposition 3) in such fashion that $f(c_3')$ becomes less than $f(c_2')$, then jumping over the critical point c_3' reunites the two components of $V(c)$ and we are back to the preceding case.

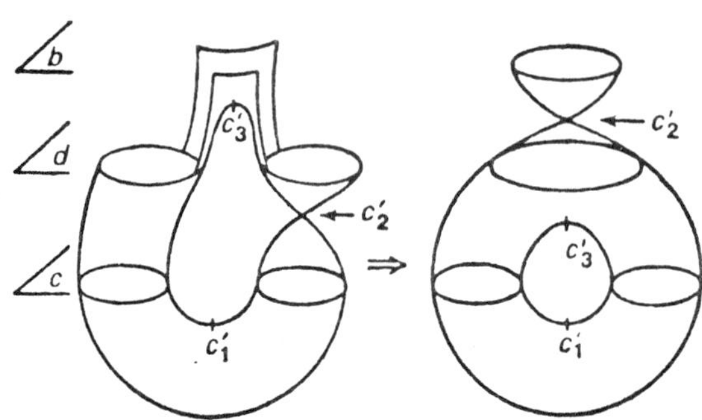

In the case of an arbitrary number q of critical points, we may assume by this procedure, that the level curves have at most two components and that for each critical point c'_j of type I, the jumping over of which decomposes a component into two, is immediately followed by a critical point c'_{j+1} which recombines these two components into one.

The general case. Assume that f has q critical points of index 1 on M. By the preceding argument, we may assume the points of type I are arranged in pairs of two consecutive points. There are pairs of the first kind and pairs of the second kind (cf. the case where $q = 2$) which shall be called, respectively, orientable and non-orientable pairs. We shall now prove the following assertion by induction on q.

If all the points are of type I and if they form orientable pairs, then M *is homeomorphic to* T_p, *where* $q = 2p$ (the critical points come in pairs), *otherwise* M *is homeomorphic to* U_q.

If all the points are of type I and all the pairs are orientable, let c be a level between $f(c_2')$ and $f(c_3')$. It follows from the study of case $q = 2$ that $M(c)$ is homeomorphic to V_1 (i.e., T_1 with a disk removed). But it follows from the induction hypothesis, by considering the function $-f$, that $M'(c)$ is homeomorphic to V_{p-1}. Hence, by gluing, M is homeomorphic to $T_p = T_{p-1} \# T_1$.

We shall now prove the second assertion. First assume that neither the two lowest points c_1' and c_2' form an orientable pair, nor that the two highest points c_3' and c_4' do. If we assume that c_1' is of type II and that d is a level between $f(c_1')$ and $f(c_2')$, then $M(d)$ is homeomorphic to U_1 with a disk removed (by case $q = 1$) and $M'(d)$ is homeomorphic to U_{q-1} with a disk removed (by the induction hypothesis). Therefore, M is homeomorphic to $U_1 \# U_{q-1} = U_q$. Now assume that c_1' is of type I, then c_2' must be also, and they form a non-orientable pair. If $f(c_2') < c < f(c_3')$, then $M(c)$ is homeomorphic to U_2 with a disk removed ($q = 2$) and $M'(c)$ is homeomorphic to U_{q-2} with a disk removed (by the induction hypothesis). Hence, M is homeomorphic to $U_2 \# U_{q-2} = U_q$.

Now, if the points c_1' and c_2' form an orientable pair, then $M(c)$ is homeomorphic to T_1 with a disk removed (case $q = 2$) and $M'(c)$ is homeomorphic to U_{q-2} with a disk removed (by the induction hypothesis). Hence, M is homeomorphic to $T_1 \# U_{q-2}$. It is easily verified by induction on $q \geqq 3$ that this space is homeomorphic to U_q. The first step in the induction was made in the case $q = 3$, where we showed that $T_1 \# U_1$ is homeomorphic to U_3.

This completes the proof.

Remark 2. In order to elucidate Theorem 4, it remains to show that there exist differentiable surfaces that are homeomorphic to T_p ($p \geqq 0$)

and to U_q ($q \geqq 0$). It also remains to show that, except for U_0 and T_0 which are homeomorphic, there are no other pairs of homeomorphic spaces among the spaces T_p ($p \geqq 0$) and U_q ($q \geqq 0$). We shall prove that in Chapter VII.

Remark 3. If the function f has a pair of non-orientable critical points of type I on M, then there exists a part of M that is diffeomorphic to the interior of a Möbius strip. It suffices to prove that for a Klein bottle ($q = 2$), there is such a Möbius strip between levels a and b (which is easy). Such a surface M is not orientable (by Proposition 3).

Remark 4. Let f be a Morse function on M which has $n(0)$, $n(1)$ and $n(2)$ critical points of index 0, 1 and 2, respectively. The proof of Theorem 4 shows that M is homeomorphic either to T_p, in which case $n(0) - n(1) + n(2) = 2 - 2p$, or to U_q, in which case $n(0) - n(1) + n(2) = 2 - q$. The uniqueness of the number p or q associated with M (Remark 2) proves that the number $n(0) - n(1) + n(2)$ depends only on M and not on the Morse function f on M. This number, denoted $\chi(M)$, is called the *Euler-Poincaré characteristic* of M. We have

$$\chi(S^2) = 2,\ \chi(T_p) = 2 - 2p,$$
$$\chi(P_2) = 1,\ \chi(U_q) = 2 - q, \text{ and}$$
$$\chi(M \# M') = \chi(M) + \chi(M') - 2.$$

§7. REMARKS ON THE GLUING OF HOMEOMORPHISMS

Let X and Y be two metric spaces, A a compact subset of X and f: A → Y an isometry. In Chapter II, §5, we defined the metric space Z glued from X and Y by f. There are two natural isometries (canonical inclusions) i: X → Z and j: Y → Z. Let X', Y', A' and f' be similar and Z' the space glued from X' and Y' by f' with i' and j' the associated inclusions. It follows from Chapter II that the specification of one homeomorphism g: Z → Z' is equivalent to the specification of two homeomorphisms h: X → X' and m: Y → Y' such that

(1) $$h(A) = A',$$

(2) $$f' \circ h|A = m \circ f,$$

and the following diagram commutes:

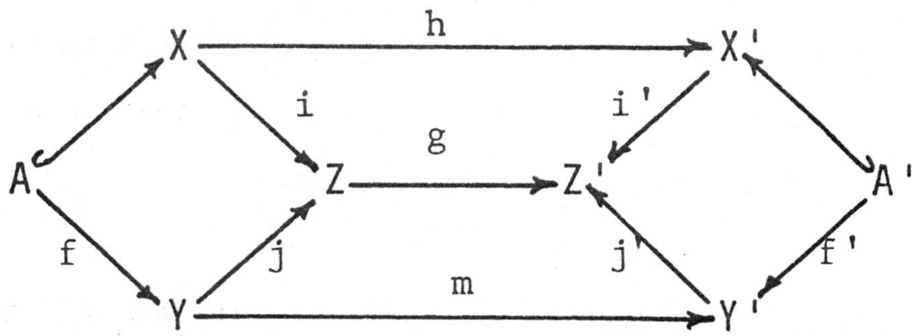

It is the theorem on the gluing together of homeomorphisms that we have used in the preceding sections in order to prove that two collages Z and Z' are homeomorphic. To this end, we have always exhibited the homeomorphisms h and m, but have not verified (1) and (2). We shall show that, *in the two cases where we have made gluings*,[1] h and m can be replaced by homeomorphisms satisfying (1) and (2).

Case 1: A *is homeomorphic to a circle.*

---

[1]This is not a general result. The conclusion depends in an essential way on the fact that one glues pieces of surfaces.

PROPOSITION 7. *Let k be a homeomorphism of the circle* S *onto itself that preserves orientation. There is a homeomorphism* K: S x [0,1] → S x [0,1] *such that*

(i) K|S x {0} *is the identity mapping, and*

(ii) K|S x {1}: S x {1} → S x {1} *is the mapping induced by k.*

Let a be the point (1,0) on the circle S whose equation is $x^2 + y^2 = 1$. We first assume that $k(a) = a$. Since k is a homeomorphism, the homomorphism $\pi(k)$: $\pi(S,a) \to \pi(S,a)$ is an isomorphism. There are two automorphisms of the group Z, the identity and multiplication by -1. We say that k *preserves the orientation* if $\pi(k)$ is the identity. We know (see Chapter I, §4) that if x is the point of S having polar angle $2\pi t$, $t \in [0,1]$, then

$$k(x) = e^{2i\pi\theta(t)},$$

where θ: $[0,1] \to [0,1]$ is a continuous function such that $\theta(0) = 0$ and $\theta(1) = 1$. Since k is injective, so is the function θ. Hence, θ is a strictly increasing function. For $s \in [0,1]$, the function $\theta_s(t) = s\theta(t) + (1 - s)t$ is a strictly increasing function of t and $\theta_s(0) = 0$, $\theta_s(1) = 1$. We then define

$$k_s(x) = e^{2i\pi\theta_s(t)}, \text{ and}$$

$$K(x,s) = (k_s(x),s).$$

Now consider the case where $k(a) \neq a$. Let $u \in [0,2\pi[$ be the polar angle of the point $k(a)$, then we have $k(x) = r_u \circ k'(x)$, where r_u: S → S is the rotation through the angle u and k': S → S is a homeomorphism such that $k'(a) = a$. We say that k *preserves the orientation* if k' preserves it. Let R: S x [0,1] → S x [0,1] be the homeomorphism defined by

$$R(x,s) = (r_{su}(x),s).$$

The homeomorphism K = R o K' works.

In the case of the gluing together of homeomorphisms, where A is homeomorphic to the circle S, we have always verified that $h(A) = A'$ and that the homeomorphism

$$k_0 = f^{-1} \circ m^{-1} \circ f' \circ h|A\colon A \to A$$

preserves the orientation. We shall show that h can be replaced by a homeomorphism h' such that $f^{-1} \circ m^{-1} \circ f' \circ h$ is the identity on A. This is possible since in the cases studied, X is of the form M(c) and A is then V(c). It follows from Proposition 1 of Chapter IV that A has a neighborhood homeomorphic to S x [0,1]. More precisely, there is a continuous injective mapping ϕ: S x [0,1] → X such that ϕ(S x {1}) = A and such that the boundary of ϕ(S x [0,1]) is ϕ(S x {0}). Let $k = \phi^{-1} \circ k_0 \circ \phi|S \times \{0\}$. It is a homeomorphism that preserves the orientation of S x {0}. Let K be the associated homeomorphism of S x [0,1] (see Proposition 7). Then $\phi \circ K \circ \phi^{-1}$ is a homeomorphism of ϕ(S x [0,1]) which coincides with k_0 on ϕ(S x {1}) and with the identity on ϕ(S x {0}). If we extend by the identity on the complement of ϕ(S x [0,1]), we obtain a homeomorphism h_0 of X onto itself (see Chapter I, Proposition 6). The homeomorphism $h' = h \circ h_0^{-1}$ is the desired homeomorphism.

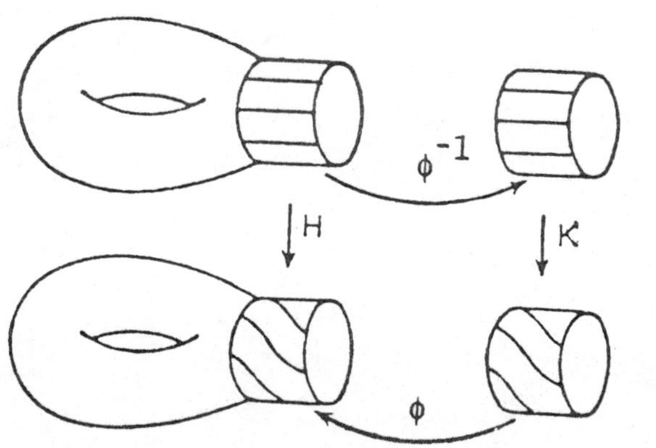

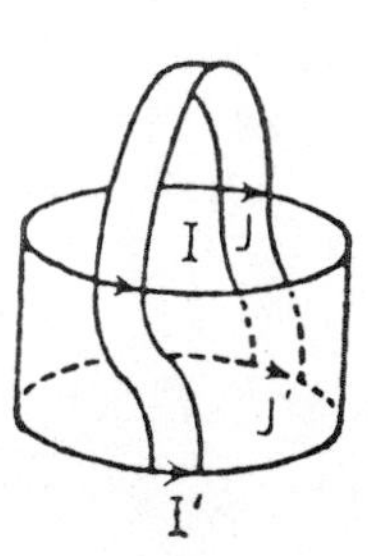

Case 2: Y is a rectangle and f(A) is formed by two opposite sides.

It is a matter of gluing together a rectangle by identifying two (opposite) sides with the disjoint arcs I and J contained in a circle V(c). We must show that, if I' and J' are two other disjoint arcs having the same orientation as I and J respectively, then there exists a homeomorphism h_0 of X onto itself such that $h' = h \circ h_0^{-1}$ satisfies (1) and (2). As before, V(c) has a neighborhood that is homeomorphic to S x [0,1]. In order to reduce the situation to the preceding case, it suffices to prove the following proposition.

PROPOSITION 8. *Let* γ, δ, γ', δ': [0,1] → S *be continuous injective mappings having images* I, J, I', J', *respectively. We assume that* I *and* J *(resp.* I' *and* J'*) are disjoint and that* I *and* I' *(resp.* J *and* J'*) have the same orientation. Then there is a homeomorphism* k: S → S *that preserves the orientation and such that*:

$$k \circ \gamma = \gamma' \text{ and } k(I) = I',$$

$$k \circ \delta = \delta' \text{ and } k(J) = J'.$$

The proof is very easy.

CHAPTER VI

KNOTS

Knot theory is at the present time rather well developed, although very complicated. It is concerned with the study of the figure formed by a simple closed curve in 3-dimensional space. That is the mathematical representation of the figures which we can form with a string. The problems posed generalize to the study of a sphere of dimension p embedded in a space of dimension n. We limit ourselves to giving some definitions and some elementary results of knot theory as examples of the utility of the fundamental group. For more details the reader can consult the book *An introduction to knot theory*, Springer, 1970, by R. H. CROWELL and R. H. FOX and the book *Knots and links*, Publish or Perish, 1976, by D. ROLFSEN.

§1. DEFINITIONS

A *knot* is the image K of a continuous injective mapping $f\colon S^1 \to R^3$. Two knots K and K' are *equivalent* if there is a homeomorphism $h\colon R^3 \to R^3$ that preserves orientation[1] and for which $h(K) = K'$. Then $h(R^3\backslash K) = R^3\backslash K'$. A necessary (but not sufficient) condition for two knots to be equivalent is that their complements be homeomorphic and in particular,

---

[1]We shall not make this concept precise.

that the fundamental groups of their complements be isomorphic. We call the fundamental group $\pi(R^3\backslash K, x_0)$, where x_0 is an arbitrary base point, the *group of the knot* K and denote it by G(K). Verifying that G(K) is not isomorphic to G(K') is a very effective technique for showing that K is not equivalent to K'. However, there are knots whose groups are isomorphic but which are not equivalent.

We shall be interested here only in *differentiable knots*. A knot K is differentiable if it is the image of a continuously differentiable injective mapping $f: S^1 \to R^3$ which has rank 1 at every point. We call f an *embedding* of S^1 into R^3. As will be seen that is equivalent to K being a submanifold of R^3. We shall assume that f is of class C^∞, which does not change anything in the matter.

With K and f as above, consider the subset N of $S^1 \times R^3$ consisting of the pairs (Θ,Y), where Y is normal to K at $f(\Theta)$, i.e., Y is normal to $Df(\Theta)$ (which is a line). We denote the obvious projection by $p: N \to S^1$, and for $r > 0$ denote by N_r the open subset of N consisting of the (Θ,Y) such that $\|Y\| < r$. Let $\psi: N \to R^3$ be the mapping $(\Theta,Y) \mapsto f(\Theta) + Y$. We can immediately verify that N is a differentiable manifold and that ψ and p are differentiable. Furthermore, ψ has rank 3 at every point of $S^1 \times \{0\}$. According to the inverse function theorem (Ch. III, §2), for each $\Theta \in S^1$ there are real numbers $\alpha(\Theta)$, $\rho(\Theta) > 0$ such that ψ induces a diffeomorphism of the open subset $p^{-1}(A(\Theta,\alpha(\Theta))) \cap N_{\rho(\Theta)}$ onto its image, where $A(\Theta,\alpha)$ designates the arc of S^1 consisting of the Θ' such that $\|\Theta - \Theta'\| < \alpha$. Since S^1 is compact we can show (by an argument analogous to that of Lemma 2 in Chapter II) that there are real numbers α, $\rho > 0$ such that, for all $\Theta \in S^1$, the mapping ψ induces a diffeomorphism of $p^{-1}(A(\Theta,\alpha)) \cap N_\rho$ onto its image.

We shall prove (perhaps shrinking ρ) that the mapping ψ can be assumed to be injective on N_ρ. Let (Θ,Y) and (Θ',Y') be two distinct elements of N_ρ. In order to have $f(\Theta) + Y = f(\Theta') + Y'$, it is necessary (by the preceding) that $\|\Theta - \Theta'\| \geqq \alpha$. We shall show that there then exists a real number $\beta > 0$ such that $\|f(\Theta) - f(\Theta')\| \geqq \rho$. If for every integer $n \geqq 1$ there were points Θ_n, Θ'_n of S^1 such that $\|\Theta_n - \Theta_n'\| \geqq \alpha$ and $\|f(\Theta_n) - f(\Theta_n')\| \leqq 1/n$, then by the compactness of S^1 the sequences $\{\Theta_n\}$ and $\{\Theta_n'\}$ would have accumulation points Θ_0 and Θ_0' such that $\|\Theta_0 - \Theta_0'\| \geqq \alpha$ and $f(\Theta_0) = f(\Theta_0')$, which would contradict the fact that f is injective. Hence, we have $\|f(\Theta - f(\Theta')\| \geqq \beta > 0$. Take $\rho < \beta/2$. If $\|Y\| < \rho$ and $\|Y'\| < \rho$, then we cannot have $f(\Theta - f(\Theta') = Y' - Y$. This proves that ψ is injective on N_ρ.

Choose r such that $0 < r < \rho$; then $\bar{N}_r$ is contained in N_ρ. We set $V_r = \psi(N_r)$ and denote the mapping $f \circ p \circ \psi^{-1}$ by π: $\bar{V}_r \to K$.

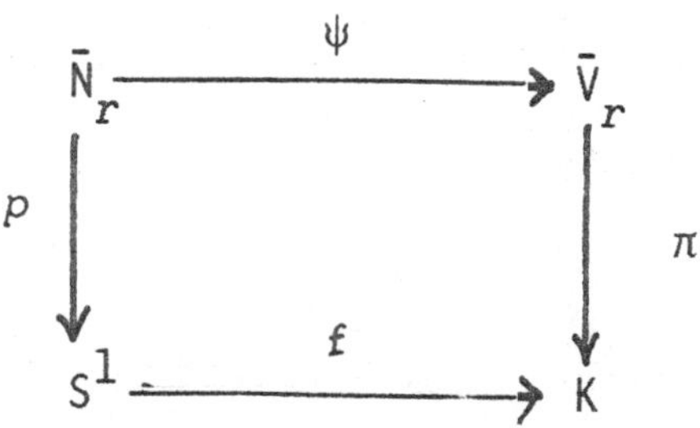

The manifold N with the map p is called a *normal bundle* of K and V_r equipped with π is called a *tubular neighborhood* of K. Now observe that we have incidentally proven that a differentiable knot is a manifold in R^3.

On the set of embeddings of S^1 into R^3 we consider the metric of uniform convergence of mappings <u>and</u> of their first derivatives (i.e., *the metric of* C^1*-uniform convergence*). An isotopy between two embeddings

f_0 and f_1 is a (continuous) path $t \to f_t$ in the space of embeddings. In other words, the mappings $t \mapsto f_t$ and $t \mapsto Df_t$ are continuous for uniform convergence. It does not suffice to say that the mapping $F(\theta,t) = f_t(\theta)$ is continuous (cf. Ch. I, §2); we also require that $D_\theta F(\theta,t)$ be continuous.

PROPOSITION 1. *Let f be an embedding of S^1 in R^3. If f' is an embedding that is sufficiently close to f under C^1-convergence, then the knots $K = f(S^1)$ and $K' = f'(S^1)$ are equivalent. Moreover, f and f' are isotopic.*

Let $d > 0$ be the distance from the compact set K to the closed set $R^3 \setminus V_r$. If $\|f'(\theta) - f(\theta)\| < d$ for every $\theta \in S^1$ then K' is contained in V_r. The mapping $\phi = f^{-1} \circ \pi \circ f'$: $S^1 \to S^1$ approaches $\mathrm{id}(S^1)$ in the C^1 sense when f' approaches f in the C^1 sense. For f' sufficiently close to f, the mapping ϕ is of degree 1 (Ch. I, §4) and its derivative (in the sense of angles) is strictly positive. Thus, it is a diffeomorphism of S^1 and $g = f' \circ \phi^{-1}$ is an embedding of the image K'. It has the special property that $\pi \circ g = \pi \circ f = f$.

For $x \in V_r$, let θ be such that $f(\theta) = \pi(x)$; we set

$$h(x) = x + (g(\theta) - f(\theta))(1 - \frac{\|x - f(\theta)\|}{r}).$$

The mapping h defined this way is a homeomorphism of V_r onto itself which leaves the points of the boundary bV_r fixed and is such that $g = h \circ f$. We have constructed h by transporting in each fiber $\pi^{-1}(\theta)$ the transformation H of the unit disk defined by $H(z) = z + a(1 - |z|)$ which, for each point b of the boundary, sends the segment $b0$ affinely onto the segment ba. Since it induces the identity on the boundary of V_r, the map h can be extended as the identity outside of V_r to a homeomorphism of R^3 (which we again denote by h), and $h(K) = K'$. By modifying h in a neighborhood of bV_r, we could make it infinitely differentiable.

We have proven the equivalence of K and K'. It is easy to see that g is isotopic to f' by deforming ϕ into $id(S^1)$ and that g' is isotopic to f by means of an affine deformation in V_r.

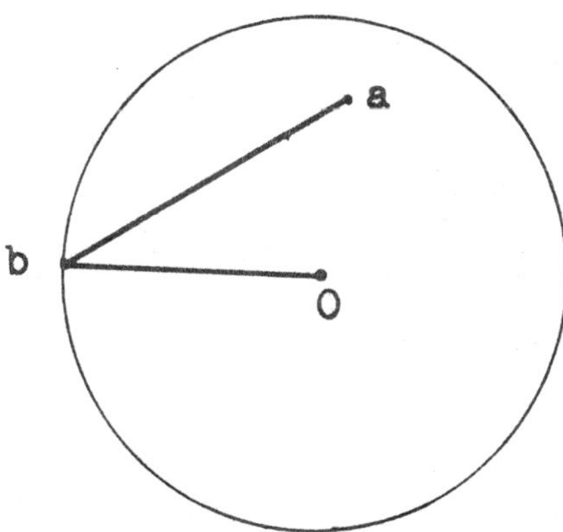

THEOREM 1. *Two isotopic differentiable knots are equivalent.*[1]

Let F: $S^1 \times [0,1] \to R^3$ be an isotopy. We set $f_t(\theta) = F(\theta,t)$, $K_0 = f_0(S^1)$, $K_1 = f_1(S^1)$ and prove the equivalence of K_0 and K_1. It follows from Proposition 1 that for $t_0 \in [0,1]$, there exists an $\varepsilon > 0$ such that the knot $f_t(S^1)$ is equivalent to $f_{t_0}(S^1)$ if $|t - t_0| \leqq \varepsilon$. The intervals $]t_0 - \varepsilon, t_0 + \varepsilon[$, where t_0 ranges over [0,1], cover [0,1]. Let $1/n$ be a Lebesgue number for this covering (Ch. II, §1). If $|t - t'| \leqq 1/n$, then $f_t(S^1)$ is equivalent to $f_{t'}(S^1)$ because t and t' are contained in the same interval $]t_0 - \varepsilon, t_0 + \varepsilon[$. From this we deduce that K_1 is equivalent to K_0 by successively making use of $K_t = f_t(S^1)$, where $t = p/n$, $p = 1,\dots, n - 1$.

The *trivial knot* is the circle $x^2 + y^2 - 1 = 0$ in the plane $z = 0$. Every circle in R^3 is obviously equivalent to the trivial knot.

[1]The converse is valid. Its proof uses difficult theorems from differential topology (see A. GRAMAIN, *Séminaire Bourbaki*, juin 1976, Lecture Notes in Mathematics, Vol. 567, Springer-Verlag, New York).

PROPOSITION 2. *The group of the trivial knot is isomorphic to Z.*

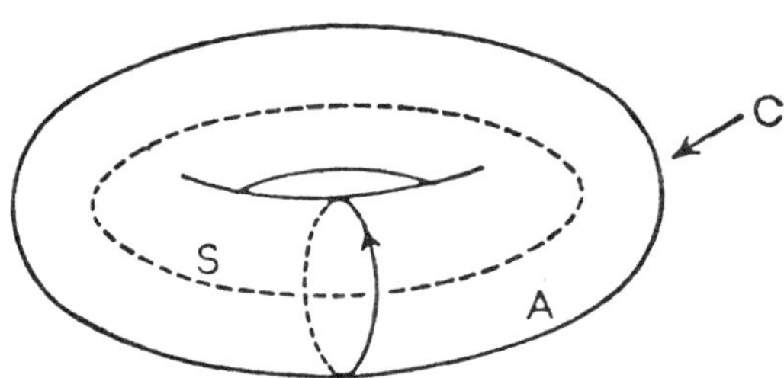

Let S be a circle in the plane $z = 0$ and let A be a solid torus with core S having the torus C as boundary. Let B be the complement of the interior of A. The inclusion $B \to R^3 \backslash S$ is a homotopy equivalence. We shall show that the fundamental group of B is infinite cyclic with generator a meridian circle of C.

We consider in R^4 the sphere S^3 which has the equation $X^2 + Y^2 + Z^2 + T^2 = 1$. Consider the stereographic projection of S^3 onto the plane $T = 0$, with center of projection the point P of S^3 such that $T = -1$. This stereographic projection is a homeomorphism $h\colon S^3\backslash\{P\} \to R^3$. The point with coordinates (x,y,z) in R^3 is the image of the point of S^3 whose coordinates are

$$X = \frac{2x}{1 + x^2 + y^2 + z^2}, \qquad Y = \frac{2y}{1 + x^2 + y^2 + z^2}$$

$$Z = \frac{2z}{1 + x^2 + y^2 + z^2}, \qquad T = \frac{1 - x^2 - y^2 - z^2}{1 + x^2 + y^2 + z^2}.$$

Consider the two subsets A' and B' of S^3 defined respectively by the inequalities:

$$(A') \qquad X^2 + Y^2 \geqq Z^2 + T^2,$$

$$(B') \qquad X^2 + Y^2 \leqq Z^2 + T^2.$$

Their intersection C' is defined by $X^2 + Y^2 = 1/2$, $Z^2 + T^2 = 1/2$; C' is the product of two circles, which is a torus. The spaces A' and B' are obviously homeomorphic. We shall show that A' is homeomorhic to a solid torus. The space A' does not contain the projection point P, and A' is homeomorphic to the space $A = h(A')$ which is defined by the inequality

$$(x^2 + y^2 + z^2)^2 - 6(x^2 + y^2) + 2z^2 + 1 \leqq 0.$$

We recognize this to be a solid torus about the z-axis whose intersection with the plane $z = 0$ is the annulus bounded by the circles having radii $\rho = \sqrt{3 \pm 2\sqrt{2}}$. Its boundary is the torus $C = h(C')$.

The fundamental group of B (the complement of the interior of A) is thus isomorphic to the fundamental group of $h(B) = B'\backslash\{P\}$. The canonical injection $B'\backslash\{P\} \to B'$ induces an isomorphism of the fundamental groups (we apply the Van Kampen theorem to $B'\backslash\{P\}$ and to a disk with center P as in Chapter II, §6, Remark 7). The fundamental group of the solid torus B' is isomorphic to Z, with a meridian of C' as generator. Therefore the fundamental group of B is isomorphic to Z, with a meridian circle of the torus C as generator.

Remark 1. One can prove conversely, that a knot whose group is isomorphic to Z is equivalent to the trivial knot (theorem of Papakyriakopoulos).

§2. THE GROUP OF A KNOT

Remark 2. A knot is said to be *polygonal* if it is the image of an injective polygonal mapping $g\colon S^1 \to R^3$. For such a mapping there exists a decomposition $0 = \theta_0 < \theta_1 < \ldots < \theta_n = 2\pi$ of the segment $[0,2\pi]$, so that on each subsegment $[\theta_i,\theta_{i+1}]$, the mapping $\theta \mapsto g(\theta)$ is affinely linear.

A differentiable knot is equivalent to a polygonal knot. Let f: $S^1 \to R^3$ be an embedding and let K be its image. If n is sufficiently large, then the polygon K' with vertices $f(0),\ldots,f(\frac{2p\pi}{n}),\ldots,f(2\pi)$ is a knot contained in a tubular neighborhood (V_r,π) of K and intersecting each fiber of π in a unique point. We could then argue as in Proposition 1 to prove the equivalence of K and K'.

Conversely, *a polygonal knot* K' *is equivalent to a differentiable knot* obtained by smoothing off K' in a sufficiently small neighborhood of each vertex.

Proposition 1 and Theorem 1 are false for polygonal knots (cf. SEIFERT and THRELFALL, *A Textbook of Topology*, comment no. 39, Academic Press, 1980). For convenience of exposition, we shall deal in this section with polygonal knots. We could verify that the use of Theorem 1 is correct by transferring the knot to a differentiable knot obtained by smoothing off the corners.

The following gives a general algorithm for obtaining the group of a polygonal knot K.

Plane representation of a knot and notation. We begin by choosing a plane P such that the orthogonal projection of the polygon K on P is "in general position", that is, it does not have either of the following properties:

a) three distinct points have the same projection; or

b) the projection of a vertex of the polygon is a double point.

Let D be a vector based at the origin and orthogonal to P. Condition b) first excludes from consideration a finite number of vectors, those parallel to the sides of K. Then b) excludes those vectors which are parallel to the finite number of planes determined by a vertex and a side of K. Condition a) excludes the vectors of a finite number of second order cones, which are the directrix cones of the quadrics containing three sides of K. Thus, we can indeed choose a suitable direction D and plane P. The projection k of K is a closed polygonal line which has as its only singularities the isolated double points which are in the interiors of the sides. For convenience we assume that P is the plane $z = 0$. Using a vertical translation if necessary, we reduce this to the case where K is situated in the half-space $z > 1$. Now we give the following graphical representation of k. Let a be the double point of k which is the projection of $A \in K$ and $B \in K$. If B is lower than A, let [B',B"] be a small segment of K having B as the midpoint. In drawing k, we suppress the projection $]b',b''[$ of the interval]B',B"[(see figure).

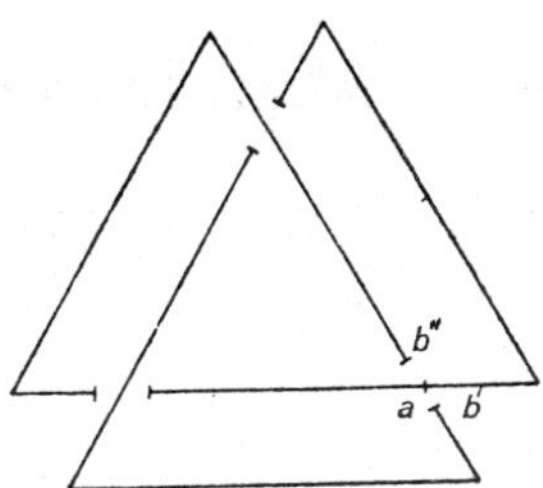

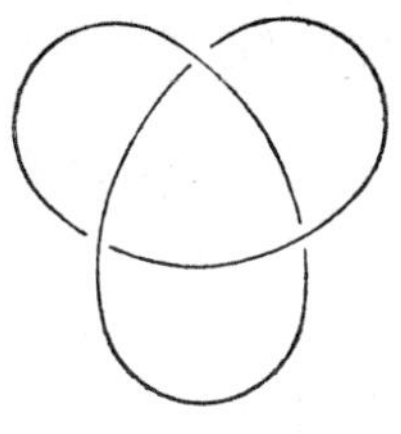

A knot K' *equivalent to* K. By isotopy (see Theorem 1), the knot K is equivalent to the knot K' obtained by replacing the segment B'B" by the union of the three segemnts B'b', $b'b''$ and b''B". Choose an arbitrary sense of direction on K and label the points B encountered in the order $B_1,\ldots,B_n$. We assume that the orientation of K coincides with that of $B_i'B_i''$ and denote by $Y_1,\ldots,Y_n$ the polygonal arcs of K which form the complement of the small segments $B_i'B_i''$. That is,

$$Y_1 = B_1''B_2',\ Y_2 = B_2''B_3',\ldots,\ Y_n = B_n''B_1'.$$

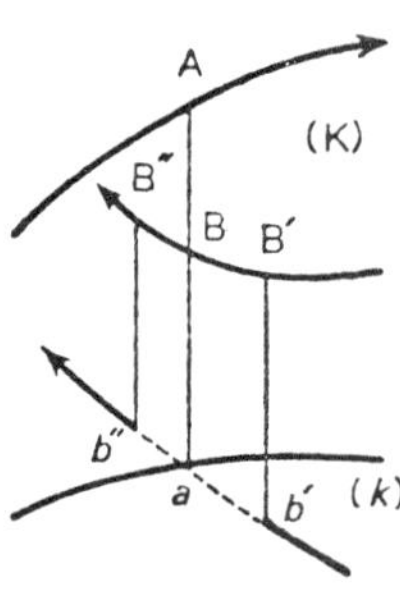

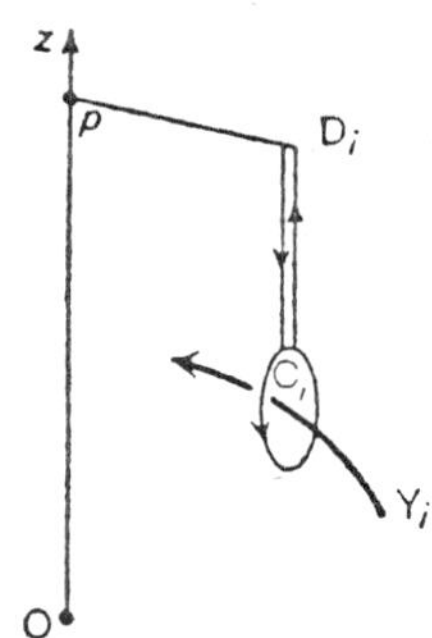

The generators of G(K). Let p be a point on the z-axis, with z-coordinate greater than that of the highest point of K. We define certain elements $x_1,\ldots,x_k$ of $\pi(R^3\backslash K,p)$. The element x_i is represented by a path γ_i obtained in the following manner: a horizontal segment joining p to a point D_i situated vertically above a point C_i of the arc Y_i, a vertical segment descending from D_i down to a neighborhood of C_i, a complete winding around the arc Y_i in the neighborhood and a return to p by backtracking on the two segments already described. The winding around Y_i is taken (with respect to the orientation of Y_i) as shown in the figure.

THEOREM 2. *The group* G(K) *of the knot* K *is isomorphic to the quotient group of* $F(x_1,\ldots,x_n)$ *(the free group on* n *generators* $x_1,\ldots,x_n$*) with respect to the normal subgroup generated by the* n *elements* $r_i = x_i x_k x_{i'}^{-1} x_k^{-1}$ *(which correspond to the* n *double points of* k*).* (Label by a_i the double point which is the projection of $A_i \in Y_k$, where Y_i and $Y_{i'}$ are the two arcs emanating from B_i.) *If* k *does not have a double point, the group* G(K) *is isomorphic to* Z.

a) We first assume that k does not have a double point. In this case k itself is a polygonal knot. Since if f: $S^1 \to R^3$ is a polygonal mapping which defines K and π: $R^3 \to P$ is the projection, then $\pi \circ f$ is an injective polygonal mapping of the image of k. Furthermore, $\pi \circ f$ is isotopic to f (isotopy along the lines of projection) and k is equivalent to K. It is easy to see that k, a plane polygon, is equivalent to the trivial knot.

b) If there are double points, we calculate the group of the knot K', equivalent to K, with the aid of the Van Kampen theorem. Consider the half-space E_1 defined by $z > 0$ and the half-space E_2 defined by $z < 1$. Furthermore, set $K_1 = K' \cap E_1$ and $K_2 = K' \cap E_2$.

LEMMA. *The group* $\pi(E_1 \backslash K_1, p)$ is the free group on n generators $x_1, \ldots, x_n$.

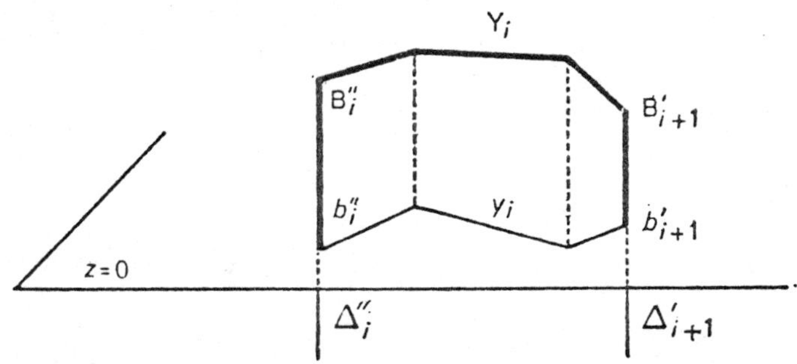

The space K_1 has n components F_i each formed by a polygonal arc Y_i (whose vertical projection y_i is without double points) with, for each of its endpoints, a vertical segemnt, $[B''_i, b''_i[$ and $[B'_{i+1}, b'_{i+1}[$. Moreover, the arcs y_i are disjoint. There exists a homeomorphism that preserves the vertical projection of E_1 onto R^2 which sends each F_i onto the union F'_i of the arc y_i and the two descending vertical half-lines Δ''_i and Δ'_{i+1} emanating from b''_i and b'_{i+1}. *The complement* $R^3 \backslash (\bigcup F_i)$ *is homeomorphic to the complement of* n *parallel lines in* R^3. We can see this by deforming, by "isotopy," the disjoint polygonal arcs y_i into parallel segments in the plane P, the half-lines Δ''_i and Δ'_{i+1} retaining the verticals of the endpoints. An argument analogous to that of Theorem 1 proves that the "isotopy" implies the equivalence for these figures as for knots. It is then not difficult to find a homeomorphism of the new $R^3 \backslash (\bigcup F'_i)$ onto the complement of n parallel lines in R^3.

The complement of n parallel lines of R^3 is homotopy-equivalent to the complement of n points of R^2. Its fundamental group is a free group on n generators, each corresponding to a complete winding around one and only one of the n lines. Rising again to $E_1 \backslash K_1$, we see that we have proven the lemma.

c) By the lemma, the fundamental group of $E_2 \backslash K_2$ is analogous because K_2 consists of n components G_i, each consisting of the segment $[b_i', b_i'']$ and two small vertical segments; we denote the generators of $\pi(E_2 \backslash K_2)$ by α_i. Finally, $(E_2 \backslash K_2) \cap (E_1 \backslash K_1)$, the complement of $2n$ vertical segments in $E_1 \cap E_2$, has a fundamental group with $2n$ generators, the β_i which encircle the Δ_i' and the γ_i which encircle the Δ_i''.

We now apply the Van Kampen theorem (Chapter II, Theorem 2). The group of K', the fundamental group of $R^3 \backslash K'$, is the product of the free group $F(\{x_i\})$ and the free group $F(\{\alpha_i\})$, amalgamated by the two homomorphisms ϕ and ψ coming from the free group $F(\{\beta_i\}, \{\gamma_i\})$. We choose the orientations of α_i, β_i, γ_i, x_i as in the figure. We then have $\psi(\beta_i) = \alpha_i^{-1}$ and $\psi(\gamma_i) = \alpha_i$. Hence, we obtain G(K') by forming the quotient of the free group $F(\{\alpha_i\}, \{x_i\})$ by the normal subgroup N generated by the elements $\phi(\beta_i)\cdot\alpha_i\cdot\phi(\gamma_i)\cdot\alpha_i^{-1}$. The relations

$$\phi(\beta_i)\cdot\alpha_i = 1 \text{ and } \phi(\gamma_i)\cdot\alpha_i^{-1} = 1$$

are equivalent to

$$\alpha_i = \phi(\gamma_i) \text{ and } \phi(\beta_i)\cdot\phi(\gamma_i) = 1.$$

From this there results an isomorphism between $F(\{\alpha_i\}, \{x_i\})/N$ and $F(\{x_i\})/N'$, where N' is generated by the elements $\phi(\beta_i)\cdot\phi(\gamma_i) = \phi(\beta_i\cdot\gamma_i)$. The loop $\beta_i\cdot\gamma_i$ winds once around the two lines Δ_i' and Δ_i'' in the same sense. It is easy to see that $\phi(\beta_i\cdot\gamma_i) = x_i x_k x_{i'}^{-1} x_k^{-1}$. Whence the theorem.

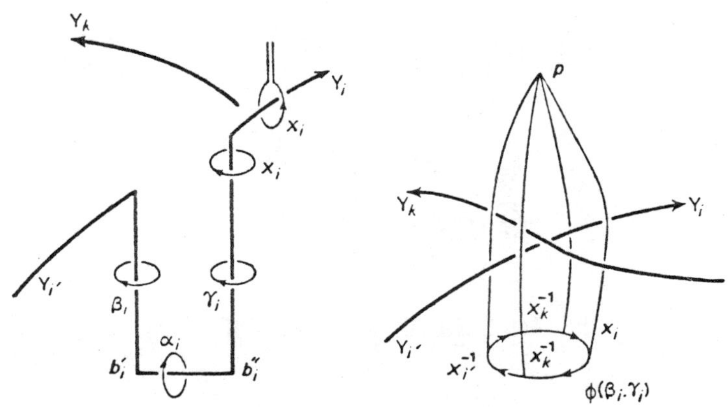

Remark 3. Any one of the relators

$$r_i = x_i x_k x_{i'}^{-1} x_k^{-1}$$

is a consequence of the other $(n - 1)$, and hence $(n - 1)$ relators suffice to define G(K). We can convince ourselves of this in the case where $n = 1$; there is only one double point, only one arc Y_1, only one generator x_1 and the relator $r_1 = x_1 x_1 x_1^{-1} x_1^{-1}$ is trivial.

Remark 4. We consider the unique homomorphism of the free group on n generators $\{x_i\}$ onto the infinite cyclic group having generator x such that $\Phi(x_i) = x$. We have $\Phi(r_i) = 1$ and Φ induces on the quotient a homomorphism ϕ: G(K) → Z which sends each generator x_i onto the generator x. We shall use this remark in the following chapter.

CHAPTER VII

SURFACES IN EUCLIDEAN SPACE R^3

If M is a compact connected surface in R^3, we shall show that M separates R^3 into two subsets. This result in fact holds for compact connected manifolds of dimension n in R^{n+1}. The proof presented here depends on the results of Chapters II, V and VI. Our proof applies specifically to manifolds of dimension 2.[1] The latter part of the chapter is devoted to examples of surfaces in R^3 and R^4.

§1. SEPARATION OF R^3 BY A COMPACT CONNECTED SURFACE

THEOREM 1 (Alexander). *Let* M *be a compact connected differentiable surface in* R^3. *The complement* U *of* M *has two arcwise connected components. Each of them has* M *as boundary and one of them is bounded.*

a) *There are at most two components*

The following argument makes use of property (P') (see Chapter III, §3) in the definition of a manifold, restricting X if necessary. Let B be the open unit ball in R^3 and let D be the open disk which is the intersection of B with the plane $z = 0$. For each point $m \in M$ there is a diffeomorphism ψ of B onto a neighborhood W of m in R^3 such that $\psi(D) =$

---

[1]For manifolds of dimension n, one uses homology theory which plays the role held here by the fundamental group, and the Alexander duality theorem, of which Remark 4 of Chapter VI is a special case.

$W \cap M$. The space $B \backslash D$ has two arcwise connected components, namely the half-balls P and N. These are the subsets of B situated respectively in the half-spaces $z > 0$ and $z < 0$. The images $\alpha = \psi(P)$ and $\beta = \psi(N)$ are the two components of $W \backslash (W \cap M)$. All the points of $\delta = \psi(D)$ are limit points of α and of β, just as the points of D are limit points of P and of N.

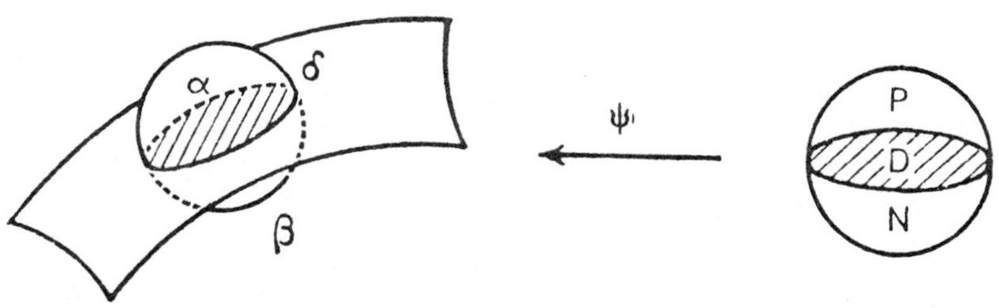

Since M is compact, the complement

$$U = R^3 \setminus M$$

is open and its arcwise connected components are open and closed in U (see Chapter I, Proposition 4). Let A be a connected component of U. Then A is open in R^3 and its boundary points (in R^3) are contained in M. There are such boundary points, for otherwise A would be open and closed in R^3 which is impossible inasmuch as R^3 is connected.[1] The set b(A) of the boundary points of A is a nonempty closed subset of M. We shall show that $b(A) = M$. If $m \in b(A)$ and ψ is (as above) a diffeomorphism of B onto a neighborhood W of m, then W intersects A, and one of the components (α, for instance) of $W \backslash (W \cap M)$ is contained in

[1]See Chapter I, Remark 1.

A. From this it follows that b(A) contains the neighborhood $W \cap M$ of m in M. The boundary of A is a nonempty subset which is open and closed in the connected space M. Hence, b(A) = M.

Therefore, a point $m \in M$ is a limit point of all the components of U. Since $W \cap U$ has only two components α and β, each contained in a component of U, the space U has at most two arcwise connected components.

b) *There are at least two components*

We take a point $m \in M$ and a diffeomorphism ψ of the ball of radius 2 in R^3 onto a neighborhood W of m such that $\psi(0) = m$ and such that $W \cap M$ is the image of those points with $z = 0$. Let S be the circle of radius 1 in the plane $z = 0$ (it is the boundary of D) and let $\sigma = \psi(S)$. Let Y be a segment of the line with midpoint m, normal to M at m. If Y is sufficiently small, one endpoint a is in the component $\alpha = \psi(P)$, and the other endpoint b is in the component $\beta = \psi(N)$, while m is the only point in common with M. Assume the contrary; i.e., $U = R^3 \backslash M$ has only one arcwise connected component. Then there is a continuous arc in U which joins a to b. Since the arcwise connected open sets of R^3 are polygonally arcwise connected, indeed there is a polygonal arc H in U joining a to b. Let K be the closed polygonal arc which is the union of H and Y. The arc K is contained in $U \cup \delta$ (the only point of K which is not in U is the point m). By a rather easy argument, we can replace K by a simple polygonal arc (without double points) which is contained in $U \cup \delta$. Furthermore, we may assume that this arc still contains the segment Y and intersects M only at m. Denote this new arc again by K.

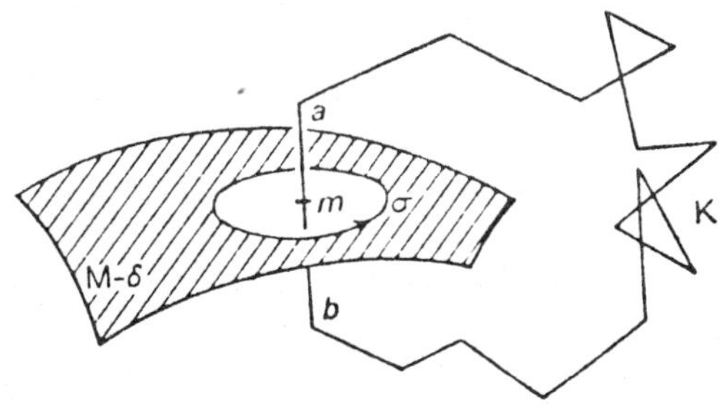

Since $K \subset U \cup \delta$, there is an inclusion

$$i\colon M \setminus \delta \to R^3 \setminus K$$

of the complements. If x_0 is a base point on the circle $\sigma = \psi(S)$, this circle is a loop in $M\setminus\delta$, we denote its class in $\pi(M\setminus\delta,\ x_0)$ by s; it is also a loop in $R^3\setminus K$, where we denote its class by s'. Then $s' = \pi(i)(s)$, where $\pi(i)\colon \pi(M\setminus\delta, x_0) \to \pi(R^3\setminus K, x_0)$ is the obvious homomorphism.

By Theorem 4 of Chapter IV, the space M is homeomorphic either to T_p or to U_q. In either case, the fundamental group of $M\setminus\delta$ is a free group (see Chapter II, Propositions 6 and 8) and the class s of σ is a product of commutators in this group if M is homeomorphic to T_p, or a product of squares in this group if M is homeomorphic to U_q.[1] It is the same for $s' = \pi(i)(s)$ in the group $\pi(R^3\setminus K,\ x_0)$ which is isomorphic to the group G(K) of the *knot* K (closed polygon without multiple points). By §2 of Chapter VI, the element $s' \in G(K)$ can be taken as a generator of G(K) because the loop σ makes precisely one winding about the edge Y of the polygon K. Let $\phi\colon G(K) \to Z$ be the homomorphism of Remark 4 of

---

[1]We shall see in §3 that there is no surface homeomorphic to U_q in R^3; but that result is based on Theorem 1. Hence, it is necessary to consider this case separately.

Chapter VI. The element $x = \phi(s')$ is a generator of Z. But, if s' is a product of commutators, $\phi(s')$ is the identity element because Z is commutative. If s' is a product of squares, then $\phi(s')$ is an even number (using the additive notation of Z). In either case, $\phi(s')$ cannot be the generator of Z. This proves by contradiction, that U has two arcwise connected components.

c) *One of the components is bounded*

Let n be a sufficiently large number so that the ball $B(n)$ (in R^3) with center at the origin and with radius n contains the compact set M. The complement of $B(n)$ is arcwise connected and contained in U. It is contained in one of the arcwise connected components of U. The other arcwise connected component of U (which is contained in $B(n)$) is bounded. Let A be this bounded component. The closure $\overline{A} = A \cup b(A) = A \cup M$ is compact.

Remark 1. A compact surface in R^3 which has n components decomposes the space into $n + 1$ components and only one of them is unbounded.

THEOREM 2. *Every compact connected differentiable surface in* R^3 *is orientable.*

By Proposition 4 of Chapter V, it suffices to show that M has a continuous field $N(m)$ of normal vectors. Let m be a point of M. There are two opposite normal vectors N_1 and N_2 of length 1 at m. One is directed into the component α and the other into the component β. The point m is a limit point of both the bounded component A of U and of the unbounded component A'. We denote by $N(m)$ that one of the vectors N_1 and N_2 which (at departure) is directed toward A'. We thus obtain a continuous field of vectors normal to M, of length 1.

§2. ORIENTABLE SURFACES CAN BE EMBEDDED IN R^3

PROPOSITION 1. *For every integer $p \geqq 0$, there is a differentiable surface* M *in* R^3 *which is homeomorphic to* T_p.[1]

We already have this result for $p = 0$ (sphere) and $p = 1$ (torus) (see Chapter III, §3, Example 2). For arbitrary p, it follows from the next lemma which we shall not prove.

LEMMA 1. *Let* M_1 *and* M_2 *be two differentiable manifolds of dimension n in* R^q. *There exists a differentiable manifold* M *in* R^q *which is homeomorphic to the connected sum* $M_1 \# M_2$.

The idea of the proof of this lemma is the following: after a translation, if necessary, of M, we may assume that M and M' are disjoint. Remove a ball from M and a ball from M' and join the boundaries thus formed by a cylinder. It remains to show that this can be done so that the result is differentiable.

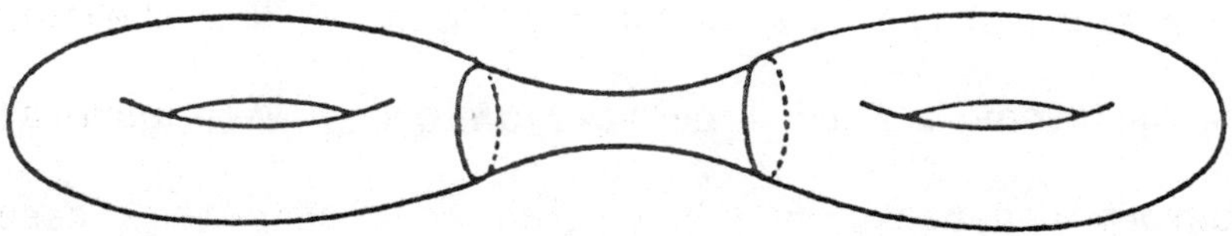

[1]Actually, two homeomorphic *surfaces* are also diffeomorphic. Consequently, every surface homeomorhic to T_p is diffeomorphic to a surface in R^3.

PROPOSITION 2. *If $p \neq p'$, the space T_p is not homeomorphic to the space $T_{p'}$.*

The fundamental group $\pi_1(T_p)$ is the quotient of a free group on $2p$ generators by a product of commutators. If we abelianize it, i.e., form the quotient using the (normal) subgroup generated by all the commutators, then we obtain the *free abelian group* on $2p$ generators Z^{2p}. Thus, if $p \neq p'$, then it is impossible for $\pi_1(T_p)$ and $\pi_1(T_{p'})$ to be isomorphic.

PROPOSITION 3. *If $q \neq 0$, the space U_q is not homeomorphic to any space T_p.*

To prove this, it suffices to show that in abelianizing $\pi_1(U_q)$ we obtain a group isomorphic to $Z^{q-1} \times (Z/(2))$. This will also prove the following proposition.

PROPOSITION 4. *If $q \neq q'$*, the space U_q *is not homeomorphic to the space $U_{q'}$.*

First, for $q = 1$, $\pi_1(U_q) = Z/(2)$ (see Chapter II, Proposition 7). Next, for $q = 2$, we have $\pi_1(U_2) = F(\alpha,\beta)/(\alpha^2\beta^2)$. If we set $\rho = \alpha\beta$, then there is an isomorphism $F(\alpha,\beta) \to F(\alpha,\rho)$ which induces an isomorphism $F(\alpha,\beta)/(\alpha^2\beta^2) \to F(\alpha,\rho)/(\alpha\rho\alpha^{-1}\rho)$ on the quotients. Now abelianize $\pi_1(U_2)$ to obtain the quotient of the free *abelian* group with generators α and ρ by the subgroup generated by ρ^2. Which concludes this case. In the general case, if $q = 2p + 1$, the space U_q is homeomorphic to $T_p \# U_1$ and if $q = 2p + 2$, the space U_q is homeomorphic to $T_p \# U_2$. The result is obtained from the preceding and the Van Kampen theorem.

§3. NON-ORIENTABLE SURFACES CAN BE EMBEDDED IN R^4 BUT NOT IN R^3 [1]

Let M be a surface homeomorphic to U_q ($q \neq 0$). By Proposition 3, M is not homeomorphic to any T_q. It then follows from Chapter V, §6 that a Morse function on M has a critical point of type II or a non-orientable pair of critical points of type I. Such a surface is not orientable (see Chapter V, §4, Proposition 5 and §6, Remark 3). Hence, M cannot be a surface in R^3 (§1, Theorem 2).

PROPOSITION 5. *For every integer $q \geqq 1$, there is a differentiable surface* M *in* R^4 *homeomorphic to* U_q.

By virtue of the lemma in §2, it suffices to prove the proposition for $q = 1$. We shall explicitly define a surface $M \subset R^4$ that is homeomorphic to the projective plane P_2.

Let $F\colon R^3 \to R^4$ be the differentiable mapping defined by

$$u = x^2 - y^2,\ v = xy,\ w = yz,\ t = zx,$$

where (x,y,z) are the coordinates of R^3 and (u,v,w,t) are those of R^4. Let S^2 be the sphere having equation $x^2 + y^2 + z^2 = 1$ in R^3 and let $f\colon S^2 \to R^4$ be the restriction of F.

LEMMA 2. a) *The mapping $f\colon S^2 \to R^4$ is a differentiable mapping of rank* 2 *at every point.*

b) A necessary and sufficient condition for two points A *and* A′ *of* S^2 *to be diametrically opposite is that* $f(A) = f(A')$.

---

[1] Let M be a manifold of dimension n. It is an important problem to determine the minimum dimension q of R^q for which M can be a manifold in R^q. In 1944, H. WHITNEY proved that every manifold of dimension n is diffeomorphic to a manifold in R^{2n}. For a *given* manifold M, the problem of calculating q was not entirely solved in 1971.

LEMMA 3. *If the mapping* $f\colon S^2 \to R^4$ *has the above properties a) and b), then its image* M *is a compact differentiable surface homeomorphic to* P_2.

Proof of Lemma 2. The Jacobian matrix of F is

$$J(F) = \begin{pmatrix} 2x & -2y & 0 \\ y & x & 0 \\ 0 & z & y \\ z & 0 & x \end{pmatrix}.$$

The determinants of order 3 extracted from J(F) are

$$2xyz,\ 2z(x^2 - y^2),\ 2x(x^2 + y^2),\ 2y(x^2 + y^2);$$

they are all zero only when $x = y = 0$. Except at the points $(0,0,\pm 1)$, the mapping f has rank 2 since F has rank 3. If $x = y = 0$, and $z \neq 0$, DF has rank 2 and its kernel is the subspace $x = y = 0$; since it is not tangent to the sphere S^2 at either of the points $(0,0,\pm 1)$, the mapping f has rank 2 at these points. Assertion b) is obvious because specifying (v,w,t) determines the line OA.

Proof of Lemma 3. According to the definition of the space P_2 (see Chapter II, §5) and property b), the continuous mapping $f\colon S^2 \to R^4$ induces a continuous injective mapping $\tilde{f}\colon P_2 \to R^4$. The space P_2 is compact (see Chapter II, §5). The closed subsets of P_2 are compact, so their images under $\tilde{f}$ are compact and hence closed in R^4. The continuous injective closed mapping $\tilde{f}$ is a homeomorphism of P_2 onto $M = f(S^2)$.

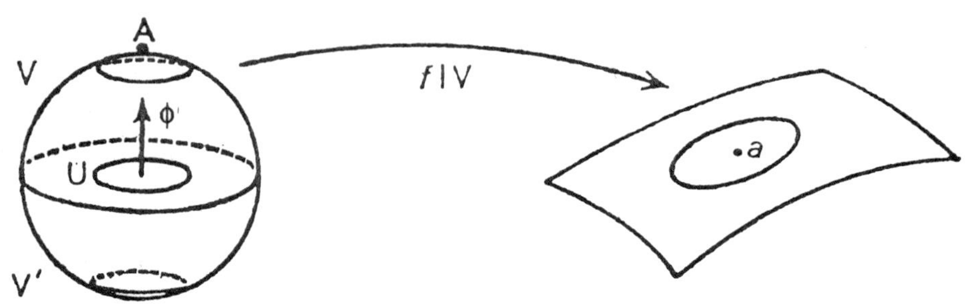

We now show that M is a manifold. Let $a \in M$ and let $A, A' \in S^2$ be the two inverse image points of a under f. Let V be an open disk with center A on S^2, with radius less than 1/2 and let V' be the disk with center A' obtained from symmetry about the origin. The image $f(S^2 \setminus (V \cup V'))$ is a compact set which does not contain a (property b)). Let W be the complement of $f(S^2 \setminus (V \cup V'))$. Then W is an open neighborhood of a and $W \cap M = f(f^{-1}(W)) = f(V)$. Let $\phi: U \to S^2$ be a parametrization of S^2 whose image is V. For example, let U be the projection of the open disk V onto the plane perpendicular to OA and containing the origin, and let ϕ be the inverse of this projection (if A is the point (0,0,1), ϕ is defined by $(x,y) \mapsto (x,y,\sqrt{1 - x^2 - y^2})$). The mapping $f \circ \phi: U \to R^4$ has rank 2 and its image $f \circ \phi(U)$ is precisely $f(V) = W \cap M$.

Remark 2. We consider the function $g: M \to R$ which associates with $a \in M$ the value $g(a) = u(x,y,z) = x^2 - y^2$, where (x,y,z) are the coordinates of one of the two points $A, A' \in f^{-1}(a)$. This function is well defined. It is differentiable. We can easily verify that it is a Morse function whose critical points are the points:

$f(0,1,0)$, index 0 (minimum), $u(x,y,z) = -1 + 2x^2 + z^2$

$f(0,0,1)$, index 1, $u(x,y,z) = x^2 - y^2$

$f(1,0,0)$, index 2 (maximum), $u(x,y,z) = 1 - 2y^2 - z^2$.

The pairs (x,z), (x,y) and (y,z) are the local coordinates in the neighborhood of each of these points, respectively.

INDEX

INDEX

INDEX

INDEX

INDEX

INDEX